D1125251

The
DRAGON SEEKERS

Other books by Christopher McGowan

The Raptor and the Lamb
Make Your Own Dinosaur Out of Chicken Bones
Diatoms to Dinosaurs
Dinosaurs, Spitfires, and Sea Dragons

The
DRAGON SEEKERS

How an Extraordinary Circle of Fossilists Discovered the Dinosaurs and Paved the Way for Darwin

Christopher McGowan

PERSEUS PUBLISHING

Cambridge, Massachusetts

A CIP record for this book is available from the Library of Congress.
ISBN 0-7382-0282-7

Printed in the United States of America.

Perseus Publishing is a member of the Perseus Books Group
Text design by Jeff Williams
Set in 11.5 point Perpetua by the Perseus Books Group
First printing, April 2001

Visit us on the World Wide Web at *http://www.perseuspublishing.com*

Perseus Publishing books are available at special discounts for bulk purchases in the U.S. by corporations, institutions, and other organizations. For more information, please contact the Special Markets Department at the Perseus Books Group, 11 Cambridge Center, Cambridge, MA 02142, or call 1–617–252–5298.

1 2 3 4 5 6 7 8 9 10 – MV – 03 02 01

To Jill, who made it happen

And Amanda, who made it work

Contents

Preface

London's Natural History Museum, like New York's American Museum of Natural History, is one of the great museums of the world. One of the reasons for its greatness is that its library houses virtually everything that has ever been published on the subject of paleontology. It also has an extensive archival section, ranging from hand-written letters and pen-and-ink sketches to sale cataloges and advertisements for lectures.

One of the earliest recollection of my fledgling career as a paleontologist is being cocooned by books in the museum's main library, poring over the writings of nineteenth-century fossilists. They beguiled me with their circumlocutory language and their passion for fossils. They enchanted me with their accounts of antediluvian creatures. And they spirited me away to their bygone world of Regency and Victorian England. I could picture myself—silk hat and frock coat—perambulating the foreshore of the Dorset coast, searching the cliffs for fossils. Or cabbing through London's horse-drawn traffic to attend a meeting of gentlemen geologists. And so the seeds were sown, unintentioned and unbeknown, for this book.

The deciding moment for putting pen to paper occurred to me six miles above the Atlantic Ocean. I was on my way home to Canada from a working trip in England, and was reading a recent publication on Mary Anning. She is the folk-hero fossilist of the nineteenth century who began making major discoveries of extinct reptiles, "sea dragons," at the age of twelve. Mary Anning has been the subject of many books, but most of these are romanticized stories for children that fail to tell the real story. What if I wrote a *serious* book about her?

There is an enormous appeal in writing a biography about the discoverer of so many exciting new fossils at such a critical period in history. This was the time leading up to Darwin's *Origin of Species,* when most English scholars were still using the Book of Genesis to rationalize the living and the prehistoric world. This was also the time of the Industrial Revolution, when England was undergoing the growth pains of changing from an agrarian to a manufacturing society.

Mary Anning's humble background, and her disenfranchised sex, made the idea all the more appealing. But why restrict a book to a single fossilist? What about her paleontological contemporaries, and all those stories of the incredible Age of Reptiles they were unearthing? Several in-depth biographies have been written about some of the key players, but nobody had attempted to bring them all together. This made no sense to me because they all lived at about the same time, they all knew each other, and they also interacted intellectually and sometimes socially. There were some remarkable characters too, including eccentrics, egotists, a workaholic, and at least one downright scoundrel. The revolutionary times in which they lived were no less extraordinary than the astonishing discoveries they were making. This was a story that had to be told.

Although I have lived most of my life in Canada, my trips back to England have been frequent, and I have visited most of the haunts of these early fossilists. Indeed, I am as familiar with Dorset's seaside resort of Lyme Regis, with its celebrated fossiliferous cliffs, as I am with my hometown in Ontario. In preparation for this book I visited some localities I had only read about before. These included some historic cave sites, made famous by the man who discovered the world's first dinosaur (though not in a cave). I am not a caver, but, thanks to the very generous assistance of a network of keen speleologists, I was put in touch with those who could help me. The last site I visited was Paviland Cave, on the south coast of Wales, some fifteen miles from the seaport of Swansea.

Elsie Little, a local caver, had arranged for one of her caving friends to take me to the spot. This was just as well because I would never have found it alone—the cave faces the sea and cannot be seen from the land. Eric, my guide, a retired college teacher, planned for us to arrive at low tide so we could climb down to the beach and approach the cave by the indirect, but safe, seaward route. We arrived at the coast shortly before noon on a gloriously bright but blustery January morning. A twenty-minute walk across fields brought us to the spot where we could clamber down the craggy rocks to the beach below. It was a fairly easy climb, but we could not get all the way down because the sea was still too high. And as we waited for the level to fall we realized, to our dismay, that the tide was still rising. A wait of several hours for the tide to go out was not an option, so Eric, an accomplished rock climber, went off to check out the direct fingernail route.

Clambering about over rocks is a regular part of fieldwork, and I am quite comfortable with heights. But when my companion returned several minutes later and pointed the way down from his elevated vantage point, I was horrified. The route led down a precipitous cliff of jagged rocks, which looked totally impossible from where I was standing. I dearly wanted to see the cave, but I also wanted to see my family again. Eric sportingly offered to return to the cave with my video camera, if I handed it up to him—a compromise I would have to accept. The climb up to Eric's level was considerably easier than it had appeared. Moreover, the route down to the cave looked considerably less daunting from my new vantage point. I therefore decided to carry on, and reached the cave without incident.

The romance of searching the rocks for fossils has enormous appeal, but few of us have had the opportunity to experience this firsthand. One year I was collecting ichthyosaurs in the rugged Rocky Mountain terrain of British Columbia. Ichthyosaurs, the extinct marine reptiles Mary Anning discovered, have been my primary research interest for most of my life. A few years before we (the Royal Ontario Museum) had discovered a whole new ichthyosaurian fauna, weathering out along the shore of a huge artificial lake. All the skeletons we had found so far were entirely new, including a small, salmon-size form I named *Hudsonelpidia,* after the nearby town of Hudson's Hope. Only one specimen existed, and I was hoping to find more skeletons of this rare beast. Imagine my elation, then, when we split open a slab of limestone along the shore to reveal a small second skeleton. Mary Anning's emotions when she discovered creatures that had never been seen before must have been equally electric.

My research for this book has taken me from abandoned dinosaur quarries to marine reptile collections. I have pored over fossilists' journals, visited their former homes, met their surviving relatives, and even examined their mortal remains. My respect for their work and veneration of their accomplishments has only been heightened by these experiences. I hope you will find their stories as compelling as I do.

Before commencing I would like to offer a word of warning, a disclaimer if you will. Just as no two people can read the same book or see the same movie and experience the same feelings, so no two people can research events of the past and arrive at identical conclusions. This book represents my readings of the past, my interpretations of recorded history, and may differ—probably in only minor ways—from other interpretations. Without the ability of traveling back in time to live the lives of the players in this narrative, we have no way of knowing who comes closest to the elusive sprite of truth.

Acknowledgments

One of the joys of writing this book has been encountering so many generous and kindly people who have helped me in so many different ways. I thank each one of you, not only for your practical assistance, which has been indispensable, but also for your enthusiastic support and encouragement. Knowing where to begin to express my deep gratitude poses a great challenge. Perhaps I should follow my paleontological proclivity and begin chronologically.

I began researching the literature for this book in the summer of 1997, at the library of the Natural History Museum. The archivist, the late John Thackray, was immensely helpful, so too were librarians Carol Gokce, Ann Lum, and Paul Cooper. Susan Snell, the present archivist, has also been very generous with her help and her time, for which I am much obliged. I gratefully acknowledge the Trustees of the Natural History Museum for permission to cite material from the NHM library. My colleagues in the Department of Palaeontology were as helpful and accommodating as always. I thank them all, with particular appreciation to Angela Milner for access to the collection and for her assistance with literature, and to Adrian Doyle for discussions of the state of preservation of the historically important Hawkins skeletons that he helped conserve. Sandra Chapman, who has a keen personal interest in the early collectors, particularly in Mantell upon whom she has published, was unstinting in providing me with literature, references, useful discussion, and her unrestrained enthusiasm. I am especially indebted to her for her support when most it was needed. I thank the Sussex Archaeological Society, Barbican House Library, for permission to quote from Mantell's transcribed journals, with especial thanks to Joyce Crow for all her help during my visit. My thanks also to C and J Clark of Street, Somerset, for valuable archival material on local collectors. Jehane Melluish has accumulated a considerable amount of information on her distant relative, Thomas Hawkins, which she freely and generously shared with me, for which I am very grateful. Michael Taylor (Royal Scottish Museum), has similarly shared his extensive knowledge of Hawkins with me, providing me with literature, references, answers to many questions, and numerous useful discussion, for which I thank him. I am very grateful to Stephen Howe and Mike Bassett of the National Museum of Wales for their generous help with archival material in their care. I thank the National Museum of Wales for permission to quote from this material, and thank Stephen for all his help in locating literature and manuscripts, and for answering so many queries. I am very grateful to Philippe Taquet (Museum National d'Histoire Naturelle), for providing me with invaluable archival information, for answering my questions, and for sharing his insights on Cuvier. I thank the Alexander Turnbull Library, National Library of New Zealand, Te Puna Matauranga o Aotearoa, for access to their valuable archival material on early English fossilists, and gratefully acknowledge their permission to quote from this. I am very grateful to their remarkably helpful staff, particularly to Tim Lovell-Smith for locating so many archival items, answering numerous questions, and for sending me material after my visit. My thanks to Stella Brecknell of the library of the Oxford University Museum of Natural History (OUMNH) for her assistance with archival material, with grateful acknowledgements to Oxford University Museum for permission to cite from this. I am especially grateful to Philip Powell, also of the OUMNH, for his generous help with literature, and for his help with so many enquiries. For permission to cite material from the Silliman Family Papers I thank Manuscripts and Archives, Yale University Library, with particular thanks to Diane Kaplan and Judith Ann Schiff. My own library at the Royal Ontario Museum has been very helpful, as always, and I would like to thank Sharon Hick, Charlotte Goodwin, Julia Matthews and Champa Ramjass. Hugh Torrens (Keele University), a walking lexicon of information on the early collectors, generously provided me with references and information for which I thank him. I also thank Liz Allan and Martin Cooke (Hunterian Museum, Royal College of Surgeons) for archival information.

Liz-Anne Bawden, the former Curator of the Philpot Museum, and Jo Draper, generously helped with information, and with ideas for illustrations, for which I thank them both. I warmly acknowledge my good friend John Fowles for countless discussions over the years on fossils, fossilists, and the Dorset coast—subjects upon which he is so knowledgeable and which are so dear to his heart. I also thank him and his wife Sarah for letting me ransack their home for illustrations of old Lyme Regis, some of which appear in this book.

Gwyneth Campling (Natural History Museum) was very helpful in tracking down illustrations and arranging to send them to me. Dennis Parsons (Somerset County Museum) kindly located the portrait of Thomas Hawkins, making it available to me, and Stephen Howe (National Museum of Wales) obtained a photographic print of De la Beche. Marion Minson (The Alexander Turnbull Library, National Library of New Zealand), obtained a digital image of Mantell's restoration of *Iguanodon*. Stella Brecknell and Philip Powell helped locate illustrations of Buckland. Clare Newman and Rhiannan Sullivan (Science and Society Picture Library of the Science Museum) helped with images of steam locomotives, and Mark Katzman (Special Collections, American Museum of Natural History) located an illustration of Cuvier. Roger Clark, Mike Taylor, and Vicen Carrió Lluesma helped me obtain a good image of the first complete plesiosaur skeleton. Wendy Cawthorne and Andrew Mussell (Geological Society) and Peter Aspen and Arnott Wilson (Edinburgh University) were tremendously helpful with picture enquiries. My sincere thanks to you all.

Before putting pen to paper I visited certain historical localities that I had not seen before. David Hill (South West Museums Service), took me to see a number of abandoned quarries in the vicinity of Street that he and Mike Taylor have been researching. I thank him for sharing these old Thomas Hawkins haunts with me. My underground explorations were made possible through an enthusiastic network of cavers, who spared no effort in helping me see some truly spectacular sights. Dave Jackson and his friend Samantha, outfitting me with overalls and miner's helmet, took me into Kirkdale Cave on a remarkably cold winter's Saturday. I thank them for generously sharing their knowledge and love of caving with me. My thanks to John Smith for his extensive knowledge of Dudley Cavern, and for showing me what remains of this astonishing underground excavation. I also thank him, and his wife Janet, for their warm hospitality. Elsie Little arranged for Eric, a fellow caver, to take me to Paviland Cave. I am deeply indebted to Eric for taking so much time and trouble in guiding me to this remarkable cave, which I could never have found, far less reached, on my own. A number of other cavers supplied me with valuable information, and with the contacts that ultimately made everything possible. My sincere thanks to Mick Day, Keith Edwards, Tony Gibbs, Graham Price, and John Wilcock. My thanks also to Tony Sutcliffe (Natural History Museum) for our many discussions on cave exploration.

I am indebted to the local collectors of Lyme and district: David Costin, Peter Langham, Chris Moore and David Sole. I thank them, not only for sharing their field experiences with me, but also for collecting the remarkable ichthyosaurs upon which I have conducted much of my research.

Kevin Padian (Museum of Paleontology, University of California), read the entire manuscript, giving me so many valuable comments and suggestions. I cannot find adequate words to express my sincere thanks for such selfless generosity from such a valued and respected a colleague. My sincere thanks also for critical readings by Polly Winsor (Institute for the History and Philosophy of Science and Technology, University of Toronto), Michael Taylor, Jehane Melluish and David Norman (Sedgwick Museum, Cambridge). Each read chapters in their own areas of expertise, giving me valuable comments and feedback.

I thank Sloan Mandel for his legal opinions during my investigation of the Hawkins affair. Thanks also to Sylvie Andrews and Sarah Stewart for French translations.

I am remarkably fortunate to have had Amanda Cook as my editor. I thank her for her unclouded vision of what this book should be, and for her clear direction in how to reach that goal. I thank her for her patience, constant encouragement, and for always being there when I needed help. It has been a very great pleasure working with her and I shall always be indebted for this valuable learning experience. Myia Williams copyedited the manuscript, taking care of so many points that needed attention, for which I am very grateful. I thank Alex Camlin for his classic cover design and Jeff Williams for designing the entire book. Marco Pavia, who coordinated production, was a constant source of help and of details on illustrations, layout, corrections and scheduling, for which I am truly grateful.

Regardless of all the help I received during the research and writing of this book, it would never have come into existence had it not been for Jill Grinberg, my agent (Anderson/Grinberg Literary Management). Her role during the formative stages of this project was critical, and if she had not been there to direct and encourage my efforts, I would have failed miserably. No words can express my gratitude.

I am remarkably fortunate in being married to Liz McGowan, whose contributions to this book are woven throughout its fabric. I thank her for enduring my long absences, even when I was at home and lost in the nineteenth century, and for always being there for me. She has been my sounding board, my cheering section, and my security blanket, and I would have been lost without her.

During my long career as a paleontologist I have been fortunate to work at the Royal Ontario Museum, and to teach at the University of Toronto. Both institutions have provided me with a supportive and stimulating environment in which to grow intellectually. During that time my researches have been generously supported by the Natural Sciences and Engineering Research Council of Canada. I would like to express my gratitude to these three organizations for giving me the best of all possible worlds.

1

In the Beginning

It is a remarkable fact that the human mind, which has had a presence on Earth for over 2 million years, only began rational thought on how species came into existence during the last three centuries. Prior to this age of enlightenment people were content with mythological explanations. These were later supplanted by formal religious beliefs, as in the Genesis account of the Creation.

Fossils, which are central to the issue of origins, have been known since the classical time of the Greeks. But it was not until the second half of the eighteenth century that the stony objects, dug from the ground, were correctly interpreted as the remains of former inhabitants of the Earth. Even then, the intellectuals who studied fossils were often unable to identify them correctly, far less to place them in their proper context. For example, the bilobed bony fossil that Robert Plot of Oxford ascribed to a giant human in 1676, and which R. Brooks labeled as *Scrotum humanum* in 1763, was actually the lower end of the femur of a dinosaur. Dinosaurs, and their reptilian kin, therefore passed unrealized, if not unnoticed, until the early part of the nineteenth century.

Why did it take so long for paleontology—and even longer for evolutionary studies—to come of age in this age of discovery? This question is made all the more perplexing when account is taken of the progress that had been made in many other branches of science by the nineteenth century. John Dalton (1766–1844), for example, derived his atomic theory of matter in 1803, Isaac Newton's (1642–1727) laws of motion were formulated over a century before, in 1687, and Edward Jenner (1749–1823) introduced immunization against smallpox in 1798. The reason for the differential development of the various branches of science probably has much to do with the graded response of the established church, specifically the Anglican Church in England. English theologians and clerics had no argument with those who tinkered with chemistry or physics, but intellectuals who questioned the biblical account of the Creation could expect the full and considerable weight of the church to bear down upon them.

The church's role in slowing intellectual progress in unraveling the remote past was much greater in Britain than in France. The French Revolution of 1789 broke the powerful grip the Roman Catholic Church had on the state, creating an intellectual milieu unfettered by religious dogma. It was in this secular environment that some of the most radical ideas of the age were spawned, such as Jean-Baptiste Lamarck's (1744–1829) ideas on the transmutation of species. But in Britain the Anglican Church was still an integral part of the establishment. Intellectuals in pre-Darwinian Britain therefore lacked the freedom of expression of ideas that ran contrary to the Bible—in the same way that today's science teachers are constrained in North American school districts where fundamentalists hold political power. The Genesis account of the Creation therefore remained the predominant view in Britain for most of the nineteenth century.

It is difficult for us in our modern world to appreciate the powerful influence the church had over philosophical and scientific issues during Darwin's (1809–1882) time. Except to many present-day Christian fundamentalists, the Book of Genesis has no relevance to the way we interpret the natural world and its long geological history. But it was not so when early fossilists attempted to interpret the remarkable creatures they discovered. Back then, the biblical account of how living things came into being was the accepted and seldom questioned truth. Charles Darwin himself records how orthodox he was in his religious beliefs when he was cruising aboard the *Beagle* (1831–1836), at least on moral issues, and there is no reason to suppose this did not extend to the Genesis account of creation too:

> . . . I remember being heartily laughed at by several officers (though themselves orthodox) for quoting the Bible as an unanswerable authority on some point of morality.

The early fossilists, like most other intellectuals of their time, recognized that the fossilized creatures they studied were no longer in existence. But others denied the concept of extinction on religious grounds. The idea that any of God's creatures had failed to survive cast aspersions on his wisdom and was thereby untenable. They argued, instead, that the supposedly extinct creatures simply lived on, undiscovered, in unexplored parts of the world. This may have been a plausible argument for small organisms, like ammonites (shelled marine animals), which could conceivably have survived in unexplored ocean depths, but, as the French anatomist Georges Cuvier (1769–1832) pointed out, this was highly unlikely for large land animals. Cuvier made a compelling case for extinction through his work on the large fossil mammals excavated near his Paris home.

Georges Cuvier was probably the greatest intellect of his age. Men like William Buckland (1784–1856), the first professor of geology at Oxford University, avidly read his published works, deferentially referring to his views on fossils, anatomy, and geology as the final authority on the subject. Cuvier's opinions were widely sought, often by sending problematic fossils to him at the Muséum National d'Histoire Naturelle in Paris, or by visiting him in person.

Cuvier pioneered the use of comparative anatomy to distinguish between species. For example, by making detailed comparisons of their skeletons, he demonstrated that the two living species of elephant (Asian and African) were separate species. Among their many distinguishing features, the Asian elephant *(Elephas indicus)* has a domed skull, compared with the African elephant's *(Loxodonta africana)* lower-crowned cranium; its tusks are more slender; it has twenty pairs of ribs instead of nineteen, and consequently twenty thoracic vertebrae (those between the neck and the pelvis to which the ribs attach) rather than nineteen; and their spines decrease in height gently toward the pelvis, rather than abruptly. These skeletal differences give the two living elephants markedly different profiles. The Asian elephant seems to have a prominent bump on its head, and its back is gently arched between shoulders and hips. The African elephant, in contrast, has a much flatter head, and the mid-back region is dished. Both species have flattened grinding teeth, and these are transversely ridged with raised enamel plates, like a coarse file. However, as Cuvier noted, there are relatively more ridges in the Asian elephant, and their edges lie parallel to one another, whereas in the African species the edges follow a zigzag pattern.

Using the same comparative methodology, Cuvier showed that the mammoths, and their relatives the mastodons, which do not have ridged teeth, were separate species from the living elephant species. Having established that mastodons and mammoths were distinct species, he argued how unlikely it was that such colossal animals could remain undetected in the world. This was especially so given that other large beasts, such as the elephant, giraffe, rhi-

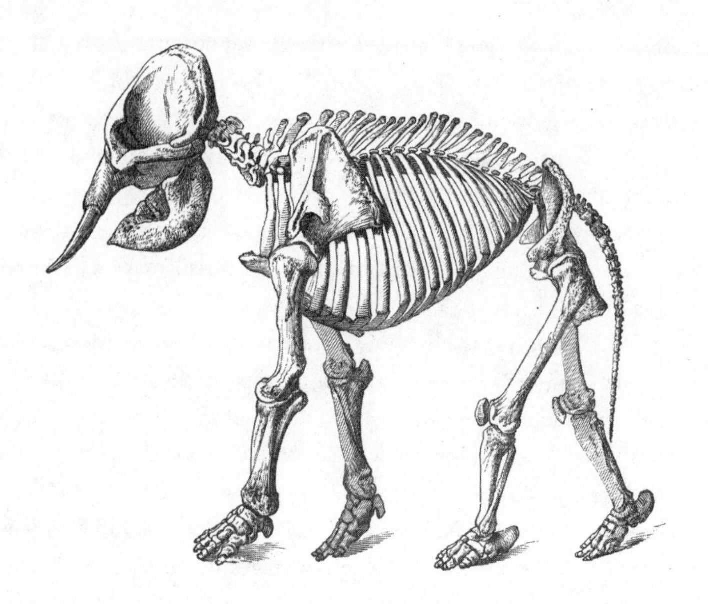

The Asian elephant has a gently arched back and a high-domed skull.

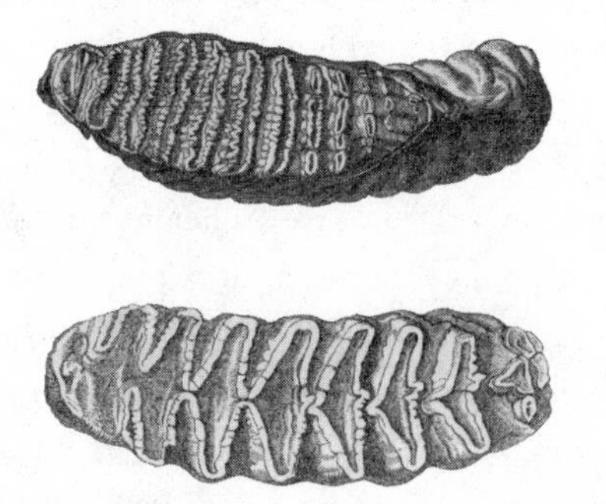

The grinding teeth of the Asian elephant (top) have more ridges than those of its African relative.

noceros, and hippopotamus, had been known since the time of the ancients. His forceful arguments for the disappearance of mammoths, mastodons, and several other large terrestrial mammals convinced most people of the reality of extinction.

Cuvier's fellow countryman Lamarck, however, disagreed. He considered that all species found in the fossil record had become transmutated into modern forms, making extinction an impossibility. According to Lamarck's transmutational theory, species were continually changing into other species. The species was therefore an artificial concept, not a natural entity. Lamarck thought the transformation came about by the acquisition of acquired characteristics from previous generations. In the often-cited example of the mechanism, the giraffe had acquired its long neck over countless generations of stretching up into trees to reach higher branches. The slight increase in neck length of each generation was passed on to the next one, thereby transforming the short-necked ancestor into its long-necked descendent. Lamarck's conjecture was the first serious attempt at a theory of evolution, though it was referred to as "transmutation" rather than "evolution"—the latter term did not come into common usage until well after the publication of Darwin's *Origin of Species*.

Cuvier, like most other intellectuals of the time, rejected Lamarck's supposition, considering that species were both real and unchanging. His opposition to Lamarckism was primarily an opposition to the idea of the transmutation of species, which denied the reality of extinction. Cuvier had little interest in the question of species transformation per se—as far as he was concerned geology had no evidence to offer on how species may have come into existence, so there was no justification in speculating on the matter.

An opportunity for the two Frenchmen to test their conflicting views on the permanency of species arose when large numbers of mummified animals were collected from some ancient tombs during Napoleon's expedition to Egypt. The age of the animals—cats, dogs, ibises, raptors, monkeys, and crocodiles—was admittedly only a few thousand years, but, during Cuvier's time, this was thought a sufficiently long period to show some indications of change. Careful comparisons showed no differences between the ancient animals and their modern counterparts, vindicating Cuvier's conviction of the permanence of species. Permanence to Cuvier, of course, meant that species did not change (evolve) through time, not that they were immune from extinction.

Cuvier also used comparative anatomy to elucidate an animal's lifestyle from its skeleton. Thus the lion's trenchant teeth, precision jaw joint, and piercing claws—the hallmarks of its carnivorous diet—could be used to identify the same traits in fossil skeletons.

But a larger intellectual challenge than interpreting lifestyles of fossil species was accounting for their present-day distribution. Fossils of marine animals were commonly found on land, sometimes at great heights above sea level, showing that considerable upheavals must have occurred during the long history of the Earth. Cuvier noted that whole groups of organisms were replaced by others during these recurring catastrophes. His studies of the sequences of fossils in the sedimentary strata had shown him, for example, that the first reptiles preceded the first mammals, and that the fossil mammals he unearthed from strata in the vicinity of Paris were different from modern ones. He also found that the proportion of modern animals increased as the rocks became progressively younger. He referred to the recurring catastrophes as *revolutions.*

Although far-reaching in their effects, he thought the revolutions were rare events, of relatively short duration, punctuating long periods of stability. These long episodes of tranquillity were evidenced by extensive formations of evenly deposited strata, such as the Chalk, a Cretaceous formation that in southern England reaches thicknesses of several hundred feet. The last of these revolutions saw the demise of mastodons, mammoths, cave bears, woolly rhinoceroses, and many other large mammals from North America and northern Europe. This last catastrophe bore all the signs of the inundation of land by water, incorporating heaps of debris and rounded pebbles into the strata. For Cuvier this revolution was no different from all the others that had wreaked havoc on the world. But for others, like Buckland, it was geological proof of the Noachian flood.

Cuvier did not think the revolutions could be accounted for by processes still operating in the modern world. He avoided speculating on what these other causes might have been, but there is no reason to think he suspected the hand of God: Cuvier, unlike Buckland, kept science separate from religion. Cuvier presented his ideas in a four-volume work entitled *Recherches sur les ossemens fossiles,* first published in 1799. This influential work, which Cuvier revised many times, was translated into several languages and was widely read. Although his radical ideas of revolutions, or catastrophes, were rejected by some intellectuals, many others, like Buckland, readily accepted them, adding his own theological spin. Catastrophism enjoyed widespread popularity for many years, but was eventually replaced by a more gradualistic view of earth history, as championed by Charles Lyell (1797–1875). Lyell sought explanations for past geological changes in terms of processes occurring in the modern world. This mechanistic explanation of past geological events is often referred to as *uniformitarianism*.

Buckland, like most other rational thinkers of his time, believed that God had directly created the world, and every living creature and plant individually. He also believed, like most, that *Homo sapiens,* made in God's own image, was the last and most special of all His acts of creation. Man occupied a special and favored place. He had dominion over the entire globe and all its inhabitants. These other creatures were placed on Earth specifically for his use and subjugation, as told in the Book of Genesis. Buckland also believed in a universal flood that had inundated the world, destroying everybody and everything that had lived on the land, save those taken aboard Noah's ark.

Although the biblical account of the Creation was broadly accepted, there was some latitude in the interpretation of the precise wording of the Scriptures. Some Christians accepted the Genesis account verbatim (as some do today), believing, for example, that God literally created the entire world in six days. But others, including Buckland, chose to take the six days as an allegory for a much longer time period. As geological knowledge progressed, greater latitudes were needed to accommodate science within the Scriptures, and a belief in the literal truth of Genesis became untenable for many. For example, it was once accepted that sedimentary (layered) rocks were formed over a period of months, as the sediments settled after the Noachian flood. However, when it was recognized that sedimentary rocks had a total thickness of many thousands of feet, such a short time frame became unacceptable. For Cuvier, Buckland, and many other thinkers, the Earth was obviously at least tens or hundreds of thousands of years old.

Buckland began his career as an ordained minister, and spent much of his academic career at Oxford documenting the geological evidence for the Noachian flood. But as new discoveries were made, even this most hallowed of biblical truths could no longer be supported by the geological record. Buckland eventually abandoned flood geology in favor of glaciation, which attributed such phenomena as the carving of valleys to the action of glaciers moving over the land. But his belief in the biblical flood was unshaken; only his interpretation of the geological evidence for the event had changed.

Buckland died in 1856, three years before publication of the *Origin of Species*. Had he lived long enough to read Darwin's account of how species had become transformed—transmutated—over long periods of geological time, he *might* just have been convinced. Darwin's arguments had, after all, persuaded many others to abandon the almost universal conviction of the permanence of species. But these minds may not have been changed if *Origin of Species* had been published many years earlier. Writing in 1876, seventeen years after its publication, Darwin recorded that "I gained much by my delay in publishing from about 1839, when the theory was clearly conceived, to 1859. . . ." He noted that the advantage of the delay had sometimes been attributed to peoples' minds becoming more receptive to the idea that species were not permanent. He himself acknowledged, in the third edition of the *Origin* (1861), the important role of earlier work in "removing prejudice" and "preparing the ground" for reception of his ideas on the transmutation of species. But he did not think this was the primary benefit. Rather, he thought the delay gave time for the accumulation of "innumerable well-observed facts . . . in the minds of naturalists, ready to take their proper places as soon as any theory which would receive them was sufficiently explained."

Much of the change in the intellectual climate in Britain is attributable to the work of early-nineteenth-century fossilists. This is not because they actively prosecuted the change—most of them, on the contrary, were firmly committed to the biblical account of the Creation and to the permanence of species. But their discoveries, by giving exciting new glimpses into the incredible world of the past, raised new and challenging questions. Consider, for example, the ichthyosaurs ("fish lizards"), a group of fishlike reptiles that were contemporaneous with dinosaurs. Their discovery raised the possibility of a link between fishes and crocodiles. Similar questions of linkage were raised with the discovery of another kind of marine reptile, the plesiosaurs. The Reverend William Conybeare (1787–1837), who coined the name "plesiosaur," thought they formed a link between modern reptiles and the more highly specialized ichthyosaurs. However, he vehemently denied this was evidence for a Lamarckian transformation of species. Instead, he considered that the linkages among different kinds of animals bore further testimony of the richness of God's creative design. Although Conybeare rejected transmutation himself, the fact that the issue was being aired publicly contributed to the change in intellectual climate. So too did the fossilists' discussions of more mundane issues, like the Noachian flood. The final defeat of flood geology, after years of public debate, ensured that this particular issue would not have to be revisited. But perhaps the greatest significance of its rejection was that it signaled the loosening of religion's grip on science.

Aside from their helping to pave the way for Darwin—even if unwittingly—the most exciting accomplishment of the early fossilists was the revelation that the Earth was once inhabited by incredible, often gigantic, reptiles. Even in our modern world of wonders, we still marvel at the latest dinosaur discoveries, so imagine what it must have been like back in Buckland's time. Those were the days when menageries stunned visitors with their first glimpses of elephants, giraffes, and other exotic creatures. When people learned that monstrous reptiles, many times larger than elephants, once roamed the land, they must have been overawed.

The early fossilists, our dragon seekers, had a rich pageant to unfold. The players were almost exclusively men, which is not surprising in those disenfranchised times. There are three principal characters besides Buckland in our narrative: Mantell, Owen, and Hawkins. Gideon Mantell (1790–1852) was a country physician who discovered some of the world's first dinosaurs. He also did much to popularize the new sciences of geology and paleontology, through his books and popular public lectures, delivered before capacity audiences. Richard Owen (1804–1892) also began his career as a physician, but left medicine to become one of the greatest anatomists and paleontologists of all time. His many achievements include the coining of the name *dinosaur.* The eccentric Thomas Hawkins (1810–1889), a man of independent but probably limited means, amassed a spectacular collection of ichthyosaurs and plesiosaurs, which he eventually sold to the British Museum. After the completion of the sale it was discovered that his specimens were not all they had appeared to be. The ensuing scandal reverberated all the way to the British House of Commons.

Besides Darwin and Cuvier, minor roles were played by Conybeare, Lyell, and Agassiz. Conybeare, close friend and confidant of Buckland, was one of the brightest and most competent geologists of his time. He was the leading authority on ichthyosaurs and plesiosaurs, and could have had a distinguished academic career. Instead, he chose to devote his life to the church. Charles Lyell, a lawyer and former student of Buckland, revolutionized the young science of geology with his *Principles of Geology,* in which he advocated uniformitarianism. Darwin took Lyell's first volume of *Principles* on the voyage of the *Beagle,* and was much influenced by what he read. In later years he would use the principle of uniformitarianism to investigate the origin of new species by seeing how selective breeding had produced such change in domestic species. Louis Agassiz (1807–1873), a Swiss naturalist and specialist of fossil fishes, is best remembered for his recognition of the Ice Age (Pleistocene), and his advocacy of the role of glaciers in bringing about physical changes in the Earth's crust that had formerly been attributed to the Noachian flood.

Among these eminent gentlemen was one leading lady: Mary Anning (1799–1847), though her lowly birth and humble station would not have earned her the appellation of *lady* in that rigidly stratified society. But in spite of, or perhaps because of, her lowly status, this remarkable woman became one of the most successful fossil collectors of all time. After discovering the world's first ichthyosaur when she was only twelve, she went on to discover the first plesiosaur and the first British pterosaur. As the supplier of many of the fossils the learned gentlemen studied and communicated to their geological circle, she played a central role in the story. However, she was often forced to share center stage with one of the gentlemen of science.

The fossilists lived in different parts of the country, but they regularly met in London, primarily at the bimonthly meetings of the Geological Society. This learned society was founded in 1807 to acquaint geologists with each other, for "stimulating their zeal," and to communicate new facts. Buckland and Conybeare both joined in 1811; Lyell joined in 1819, Mantell in 1820, and Hawkins in 1832. Darwin became a member in 1836, on returning to England from his world voyage aboard the *Beagle.* Owen joined the following year. Anning, being a woman, would not have been allowed to join even if she had wanted to, but she was made an honorary member after her death. The geological circle exchanged their ideas through reading and discussing their scientific papers. They dined together, and occasionally visited each other's homes. Some went fossilizing and geologizing together, and some even accompanied one another on Continental excursions. But they did not all get on well together, and there were all the usual rivalries and jealousies that mark our modern human interactions.

No less extraordinary than the fossilists themselves were the times in which they lived. Their story begins in England, during the late teens of the nineteenth century, and ends just before publication of the *Origin of Species.* These were times of great political unrest, when the turmoil in France, which did not end with the Revolution, seemed destined to sweep across the English Channel. This was also the time of the Industrial Revolution—the clanking behemoth that rumbled across Britain like a steam locomotive, changing lives and fortunes at a rate unprecedented until our modern age. It was a time, too, when the rich, who alone could afford the luxury of holidays, discovered the restorative benefits of the sea and sea air. Among the fashionable seaside resorts that sprang up along the coast to cater to their needs was Lyme Regis, where Mary Anning lived.

2

Dragons by the Sea

Lyme Regis today, on England's picturesque Dorset coast, is a typical seaside resort—all buckets, spades, and ice cream, beaches, cliffs, and crowds. The cries of gulls and the sound of surf are the familiar accompaniments of the sea, but one sound that seems strangely out of place is the ring of steel on stone. The sound of geological hammers striking rock has become the *chorus familiaris* along this stretch of the coast, however, come summer and winter. It is a sound that has rung out for two hundred years.

People were first attracted to the sea cliffs of Lyme Regis for the limestone, primarily quarried for making lime and cement. It was also used as a building stone, though its rapid weathering made it unsuitable for exterior use. There used to be a kiln on the foreshore, just to the west of the town, for converting the limestone into quicklime, the first step in the manufacture of cement. The quarried stone was trundled along the beach in handcarts, and rusting remnants of the rails can still be seen today, partly buried beneath the shingle.

The cliffs to the east and west of Lyme are multilayered, like a giant gâteau, limestone ledges alternating with bands of shale. The entire sequence of shales and limestone ledges is called the *Blue Lias*—blue for the blue-gray color of the limestone, and lias, probably from the Gaelic word for a flat stone. The Blue Lias belongs to the lowest division of the Jurassic—the middle period of the Mesozoic Era. The sediments are exclusively marine, and were deposited on an ancient seabed, close to land, between 195 million and 210 million years ago. The shales, a darker shade of gray than the limestone, are called paper shales because their thin layers resemble the pages of a closed book. And, like the pages of some ancient tome, sealed shut by aeons of time, they can be pried

The Blue Lias cliffs of Lyme Regis, showing alternating bands of limestone and shale.

apart. The pages are usually blank, but sometimes they bear the remains of fossils. The early quarrymen referred to the fossils as curiosities, and had no knowledge of their true origins. They thought they were formed by some exotic processes, as when lightning struck the rocks to form "thunderbolts." This was the name they gave to the bullet-shaped shells of belemnites, extinct relatives of the modern squid.

Among the most common fossils at Lyme Regis are ammonites, the coiled shells of another kind of extinct mollusc, again related to the squid. Once called "snake stones" because of their resemblance to a coiled serpent, they were sometimes carved to give them the head of a snake. But the ammonites are often squashed flat in the shales, like pressed flowers, making them less valuable as collectibles. Ammonites come in a variety of sizes, from the thumbnail sketches etched on the paper shales to the massive cartwheels, abandoned at the foot of the cliff to the west of the town. The cliffs along this stretch of the coast are highly fossiliferous, making this one of the richest fossil localities in Britain. This explains why these four miles of coastline have been such a powerful magnet for fossil hunters.

Lyme Regis became a popular tourist resort during the early days of the nineteenth century. The view from the town, looking toward the east.

Sometimes the quarrymen collected the curiosities they found, especially the less common ones. These included "verteberries," as they called isolated ichthyosaur vertebrae. But most of the fossils probably passed unheeded, only to be destroyed by the hammer or the kiln. This was because nobody seemed particularly interested in fossils or understood what they meant in those days.

That began to change during the first part of the nineteenth century, for two reasons. First, there was an increasing awareness among intellectuals that fossils were the remains of organisms that once lived on the Earth, in a world quite different from the present one. Second, there was an economic interest, occasioned by the growth of Lyme Regis as a seaside resort. Hitherto, the wealthy had flocked to spas like Bath for the supposed medical benefits of natural mineral waters. Now it became fashionable to take to the sea, not to swim and sunbathe, which were pleasures still to come, but to immerse one's body in the briny for the supposed beneficial effects to the constitution. The bathing machine—a wooden contraption on wheels that was drawn to the water's edge by horse—was invented so genteel souls could take to the sea with the modesty and decorum required of the times. Lyme Regis became one of the

Mary Anning, with her dog Tray, painted sometime before 1842.

more fashionable seaside resorts, with luminaries like Jane Austen among its visitors. The curiosities found in the cliffs now became popular souvenirs, making it worth a quarryman's while to collect fossils to sell to the rich visitors. Lesser fossils, like belemnites and ammonites, passed hands for mere pennies or shillings, but an ichthyosaur, even an imperfect specimen, would

A contemporary sketch of Mary Anning in the field.

fetch many pounds. A few of the residents became fossil collectors, both to serve the visitors' needs for keepsakes and to supply the discriminating collectors and intellectuals. But none were as successful as Mary Anning.

By 1820 Mary Anning had already earned a reputation, if not a comfortable living, from her fossiling. Neither pretty nor plain, she probably looked more mature than her twenty-one years suggested, but hers had not been an easy life. She went fossiling to put bread on the table, and had been a professional collector since she was eleven. But economic necessity went hand-in-hand with a natural curiosity. She read everything and anything she could acquire on fossils, even though she was often able only to borrow the original publications. In these instances she frequently copied out the entire article, including illustrations, and there are several examples of these in the archives of London's Natural History Museum.

One in particular so haunted me that I obtained a xerox copy. I have it in front of me as I write, alongside a copy of the original. It is a paper on marine reptiles, written in 1824 by William Conybeare, arguably the brightest geological light in England at that time. There are eight full pages of illustrations, and I am hard-pressed to distinguish the original from the copy. With meticulous care and attention to detail she faithfully copied each illustration, even taking the trouble to label every plate with the full citation of the journal. And, on a separate sheet, is a sketch of a recumbent terrier, chin on floor, eyes open, the hint of a canine smile on its face. Likely this was penned as the faithful dog watched his mistress transcribe the article.

She probably went out searching the rocks most days, summer and winter, fair weather and foul. There was usually little to show for her efforts come day's end, but that was the nature of her work. Except in the tourist season, she probably had the beach to herself most days, and would have been completely alone during the winter months. Although winter temperatures usually stay above the freezing point along that stretch of coast, it can be bone-chillingly cold in the damp sea air. By all accounts she dressed for comfort rather than fashion, and must have looked a sartorial enigma in her battered top hat and cloak, with multilayered skirts to keep out the cold.

The common fossils, the animals without backbones, or invertebrates, were her stock in trade for the summer business. But they did not fetch much money. She recalled selling her first specimen, possibly an ammonite, to a lady fossilist for half a crown (one-eighth of a pound). Her first major find, a large ichthyosaur skull, was sold for the princely sum of £23. But vertebrate fossils are much rarer than invertebrate ones: they usually comprise a much smaller portion of the fauna, and they are less likely to be preserved than hard-shelled animals like ammonites and belemnites. Difficulties aside, she knew there were unimaginable creatures slumbering there in the rocks, just waiting to be discovered. You had to *think* fossil, *feel* fossil, *believe* fossil, to find them.

She certainly knew her business well.

"I once gladly availed myself of a geological excursion . . . with Mary Anning," recalled a visiting gentleman, "and was not a little surprised at her geological tact and acumen. A single glance at the edge of a fossil peeping from the Blue Lias, revealed to her the nature of the fossil and its name and character were instantly announced."

Although she had an eye for fossils, she could not find them until they had been exposed by weathering—an achingly slow process. But when wind and rain and frost and sun had done their work, she would find them, peeking through the surface. Others were buried so deeply in the cliffs that it would be aeons before they were ever discovered.

In 1820 she was still living with her mother, also named Mary, and her brother Joseph, three years her senior. The three lived in a tiny cottage in the

The view toward the west from Lyme Regis. Notice the bathing machines along the beach.

center of the town, overlooking the sea. She hoped to have a proper fossil shop some day, with a glass-fronted window to display her wares, but for the time being she had to be content with a table set up outside.

It is often forgotten that her mother was a collector in her own right, and had been running the family fossil business for a decade or more—certainly since the death of her husband, Richard, in 1810. One reason for the oversight is that, in sharing the same name, the mother is often mistaken for the daughter. Joseph was also a collector, taking an active role in the family business, at least until the early 1820s, when his job as a furniture upholsterer left no more time for fossils. Indeed, it was actually Joseph who found the world's first ichthyosaur, probably late in 1810 when he was fifteen, though it is usually his sister, who collected it the following year, who is solely credited with the discovery.

Mary Anning's fossil excursions along the shore would have taken her either east or west of the town. The western route would have taken her along the Walk (now called Marine Parade), with its bathing machines, and past the

Cobb. This massive stone wall, which curves out to sea like a beckoning finger, has protected the harbor and town of Lyme Regis since the Middle Ages. A good winter storm can hurl cascades of seawater high above its solid ramparts, but a summer sea can be all blue innocence, lapping gently at the stone fabric on a fitful breeze. Beyond the Cobb is Monmouth Beach, and beyond that, out of sight around the point, lies Pinhay Bay. The walk to Pinhay would have given her some two miles of Blue Lias to explore.

A journey in the opposite direction, to the east of the Cobb, would have brought her to a broad reef of shale, low on the beach, right by the town. Known today as the saurian shales, it is where many of her reptilian fossils were collected. It is also where modern collectors have found most of the ichthyosaurs. The best time to prospect the reef is on a low spring tide, which comes twice each month. The water is then at its lowest ebb, so more reef is exposed. And the best time to search is after a gale, when most of the seaweed has been scoured from the rock.

The cliffs closest to town comprise crumbling black shales and marl—a fine-textured clay containing limestone nodules. Ammonites are the most common fossils in these cliffs, and are often found in the nodules. Cracking open a nodule with a sharp tap of a hammer can be like playing the three-shells game. There might be an ammonite inside, maybe even part of an ichthyosaur, but most of them turn out to be barren.

A little farther on toward the east brings a walker to Church Cliffs, and the familiar layer-cake alternation of limestones and shales of the Blue Lias. It still is a favorite spot with the summer fossilists because of its closeness to town, and Mary Anning did much of her collecting there for the same reason. Less than half a mile farther on is Black Ven, a prominent feature rising to a height of over 400 feet. It lies immediately west of the small village of Charmouth.

Black Ven is a brooding, foreboding prominence, especially on a sunless winter's day. Small rivulets of water, less often seen than heard, trickle and tumble down its soft, crumbling face. Even on the driest of summer days there is some water, somewhere. The dry marl is as hard and firm as rock, but when it rains water cascades like mountain streams, transforming it into a soft and glutinous mass. This makes the cliff prone to slump, especially during the winter months. Cliff falls occur without warning, making Black Ven a particularly hazardous place to collect fossils. Anning herself narrowly escaped death during one such fall, but her dog, Tray, was not so lucky, and was buried alive. The entire stretch of cliffs, from the marls at the foot of Black Ven in the east to the Blue Lias of Pinhay Bay in the west is prone to landslides, sometimes with catastrophic results. Groundwater seeps into the dry shales and marls and turns them into a liquid slurry that can no longer support the burden of the overlying rocks. There have been some impressive landslides over the years, when whole sections of land have slid toward the sea. Regardless of the inherent dangers, landslides often reveal hidden fossils, attracting collectors like birds behind the plough.

Fossils were Mary Anning's salvation, and not simply because they kept her fed and clothed. Fossil hunting provided intellectual sustenance as well as a sense of self-worth. Everything was set against her: her sex, her parents' low social rank, their poverty, and the fact that her father was a Dissenter, one who did not accept the doctrine of the Anglican Church.

Women were subordinate to men in all aspects of life. Only men could vote (provided they earned enough money), attend university, or hold public office or any other responsible job. Among the few jobs considered acceptable for women of the lower classes was domestic service and farmwork, together with the new manual jobs in recently opened factories. Women of more breeding had even fewer choices, with little else beside teaching and acting as ladies' companions. Society was rigidly structured along class lines. Although it was possible to rise above one's station through the acquisition of wealth, thereby acquiring some degree of respectability among the ruling classes, low birth *and* lack of wealth were passports to obscurity.

The church, specifically the Anglican Church, was a dominant influence in all affairs, and most souls regularly attended church on Sundays. The Dissenters attended their own places of worship, drawing the scorn of the more conservative Anglicans. Indeed, the least compromising Anglicans discriminated against Dissenters, and it is likely that the Annings experienced some disadvantages in a small and conservative town like Lyme Regis. And into this world came an intelligent female with an enquiring mind, a will of her own, and a free spirit.

After visiting Lyme Regis in 1824, one Lady Silvester wrote about Mary Anning:

> The extraordinary thing in this young woman, is that she has made herself as thoroughly acquainted with the science. . . . It is certainly a wonderful instance of divine favour—that this poor, ignorant girl should be so blessed, for by reading and application she has arrived to that degree of knowledge as to be in the habit of writing and talking with professors and other clever men on the subject, and they all acknowledge that she understands more science than anyone else in this kingdom.

Indeed, Anning was far more than just a collector of fossils. She analyzed her finds, often comparing the anatomies of the fossils with those of their living relatives. In one case, she obtained two specimens of the cuttlefish, a close relative of the squid, to compare with fossils of cuttlefish from the Blue Lias. One of them was still alive, and she wrote that, "whenever it was touched [it] ejected a purple fluid. . . . " This fluid, called sepia (from the generic name of the cuttlefish), was once used as a watercolor. Remarkably, Miss Elizabeth Philpot, a wealthy Lyme collector who befriended the young Mary Anning, extracted some sepia from a liassic fossil and used it to paint some stunning illustrations of her own ichthyosaurs. Anning dissected the second cuttlefish specimen, observing that it had an "ink bag exactly resembling the fossil one . . ." as well as "a second small bag (like the gisard [*sic*] of a fowel [*sic*]) containing a number of horny triangular little pieces (I should think to assist its digestion) . . . ." This was an astute observation of the buccal cavity (mouth), which has minute horny teeth for breaking up food. She also dissected a fish, probably a skate, for comparison with an unusual fossil fish she had found.

As time passed and she learned more about the fossils she collected, her knowledge surpassed that of many gentlemen fossilists. However, it was always they who formally studied her specimens, publishing their findings in learned journals, often without even mentioning her name. This would not have been an issue with Anning during the early years, as she struggled in relative obscurity. But as her discoveries won her acclaim and celebrity and as she grew intellectually, she naturally became resentful. It has been claimed that William Buckland, the Oxford geologist, was a target of her scorn, as recorded by a contemporary diarist: "Mary Anning's knowledge of the subject is quite surprising—she is perfectly acquainted with the anatomy of her subjects, and her account of her disputes with Buckland, whose anatomical science she holds in great contempt, was quite amusing."

Unfortunately, we are not privy to what these disputes might have been. Anning did not keep a journal, and the relatively few manuscripts that have survived, mostly letters, make no mention of any discord with Buckland, or any other gentleman. Indeed, her letters to Buckland, a frequent visitor to Lyme whom she had known for many years, are most cordial. For example, when Buckland asked her to make some measurements on the depths of wells, she told him she was "agreeable" to the request. She sent him numerous measurements, which must have put her to a great deal of trouble, concluding: "Mother joins me in best respects to Mrs. Buckland and the dear Children. Trusting you will think the specimen worth the carridge [*sic*]. With grateful

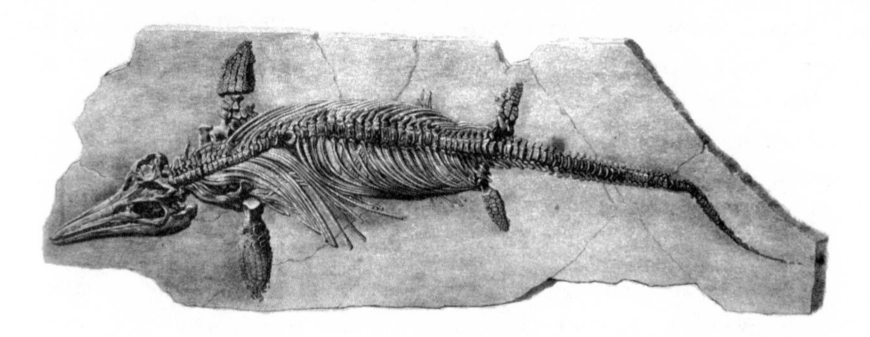

A contemporary drawing of *Ichthyosaurus communis,* the most common ichthyosaurs from the Blue Lias.

thanks for the kind offer to dispose of the skeleton for me, I remain Sir your obliged humble Servant, Mary Anning."

Although it could be argued that she was best served by remaining on good terms with him, if she did have any negative feelings toward him, she did a very good job of hiding them from the inhabitants of Lyme too, where: "local gossip preserved traditions of his adventures with . . . Mary Anning, in whose company he was to be seen wading up to his knees in search of fossils in the blue lias. . . . "

The year 1820 was a singularly unproductive one for Anning, in which she found no fossils of any consequence. It would have been cold comfort for her to think of her previous paleontological successes at that time, but her past accomplishments were quite extraordinary. Not only had she collected the world's first ichthyosaur, or fishlike reptile, but Anning had collected several more complete specimens, enabling the geological gentlemen to obtain a clearer picture of ichthyosaurian anatomy. In this way, the world at large was beginning to get a picture of what sort of creatures inhabited the seas during the remote antediluvian (pre-flood) past. In marked contrast to the modern seas that Mary Anning knew, where fish were the dominant vertebrates, those ancient waters teemed with reptiles. This must have struck her, and the other fossilists, as a remarkable fact. But what sort of animals were these sea dragons she had discovered?

Ichthyosaurs are extinct reptiles, so highly adapted for living in the sea that they bear a superficial resemblance to fish—hence their name, meaning "fish lizard." They ranged in size from a salmon to a rorqual whale. The skull has long pointed jaws, filled with sharply pointed teeth. But the most prominent feature of the skull is its enormous orbit (eye socket). The forelimbs and hindlimbs are highly modified as fins, and the individual bones, save for the upper ones (humerus and femur) have lost the usual shapes associated with legs and fingers. Instead of being elongate, these bones are compact and closely packed together, like the individual tiles of marble in a mosaic. And, unlike almost every other animal, there are usually more than five fingers in the hand. The shoulder and pelvic girdles, to which the forelimbs and hindfins attach, are relatively small and not very obvious. The neck is so short that the forefins lie almost immediately behind the head, and the vertebral column (backbone) gently arches between the shoulders and hips.

An unusual feature of the vertebral column is that it is kinked downward in the tail region. The early fossilists thought this was a postmortem effect, and it was not until some especially well-preserved skeletons were discovered in Germany during the late 1800s, in which the body outline was preserved as a thin film of carbon, that fossilists realized the kink was natural. The downturned kink formed the lower lobe of a crescent-shaped tail. Sharks have a somewhat similar tail, but their vertebral column is kinked upward.

The vertebrae—the individual bones of the vertebral column—comprise thin disks of bone, like tea-biscuits, with a depression in each round face. Each one forms the main body, or centrum, of the vertebra. Each centrum was attached to a neural arch, a V-shaped structure through which the spinal cord passed, when the ichthyosaur was alive. The neural arch continues as a blade-like neural spine. When we run a finger down our backs, the knobs we feel are the tips of our own neural spines. Ichthyosaurs were unusual among reptiles in not having their neural arches fused (permanently attached) to their respective vertebral centra. As a consequence, the more robust, hence more readily preserved, centra are often found as fossils—the quarrymen's verteberries. The ribs extend all the way from the shoulder region to the kink in the tail. Those in the chest region are long and gently curved, rapidly decreasing in length beyond the pelvis.

The first ichthyosaur remains were found before Mary Anning was born. But timing, as history has taught us, is everything, and their significance was lost on the learned gentlemen of the time. Instead of recognizing that the skeletal remains were unique, these less well-informed gentlemen generally dismissed the finds as belonging to some sort of crocodile. Fortuitously, Anning happened to be on the scene at a time when there were anatomists sufficiently enlightened to realize that the large crocodile-like skull she had collected represented an important new kind of animal. Even then, it took some time before the ichthyosaur's true affinities were realized. Sir Everard Home, a surgeon who first studied Mary Anning's specimen, initially thought it was a new kind of crocodile. Then he thought it was a fish, though by no means "wholly a fish . . . ," but one that formed a connection between fish and crocodiles. Five years later he changed his mind, concluding it to be an amphibian, intermediate between salamanders and lizards. It was not until 1821, a decade after Anning's discovery, that ichthyosaurs were finally recognized as being reptilian.

Undeterred, if dispirited, by her lack of success in 1820, Anning probably went out prospecting most days, coming back with ammonites and belemnites, brittle stars, shells, and sundry other fossils, but nothing that would command a high price. Then, in the winter of 1820–1821, she found something in the rock that must have shaken her to the core, something that was entirely new to science. We do not know exactly where or when she found the specimen, but it was in a soft marl and might have been found in the saurian shales on the beach. Nor do we know what first caught her eye, but the chances are that only a small part of the skeleton was initially exposed. She might have seen some of the terminal bones of one of its four limbs. If this had been the case she would have been astounded, because the finger bones are reminiscent of our own. Perhaps she thought she had found the remains of some ancient human.

If the skeleton had been found in a hard sediment she would probably have gone for help—we know from contemporary accounts that she used to employ locals to assist her with the larger collecting jobs. But the matrix was so soft that the specimen would have been easy for her to collect alone. She probably began by scraping away the overlying marl, no doubt thinking she had found only a few scraps of a skeleton, as was so often the case. But the bone just kept on going. She would eventually realize that she had found an almost complete skeleton. Within an hour or two she would have exposed all that remained of the bizarre creature.

There was no skull, which was a great pity, but most of the rest of the skeleton had been preserved. It was so entirely different from any other creature she had seen that she would have known, beyond any shadow of doubt, that she had discovered a new kind of sea dragon. Instead of the ichthyosaur's piscine fins this animal had long slender paddles, and these were attached to large platelike girdles, covering much of the chest and abdominal regions, like fore and aft breastplates. The paddles had five fingers and toes, comprising long slender bones, much like our own. Another singular feature was the creature's remarkably long neck—as long as its body and longer than its tail. There were many other unusual features, but she was probably more anxious to transport it to the safety of her workshop than ponder its anatomy on the foreshore.

The modern way of collecting a fossil skeleton is to strengthen all the exposed bones with a plastic varnish. Next, a trench is cut around the periphery of the specimen, using a hammer and chisel, sloping this inward to undercut the block. The bones are then covered with tissue, to stop them from sticking to the plaster and the burlap layers that are used to cap the block. The undercut on the block locks the cap in place once the plaster sets. The block is then cut free from the underlying rock, and flipped over. A second burlap and plaster jacket is applied, overlapping the first, completely encasing the specimen in a tough protective layer. The whole thing is referred to as a field jacket. If the specimen is too large to be taken out as a single block, which is usually the case, it is divided into a number of contiguous ones. Before each block is made up into a field jacket, a sketch is made of the way they fit together, so they can be reassembled in the laboratory. This ensures the continuity of the skeleton. The use of field jackets, developed by the dinosaur hunters of the American West, dates back to the late nineteenth century.

The fossilists of Mary Anning's time did not strengthen the bones, nor did they use any plaster. They simply chiseled out the slab of rock with its contained skeleton as best they could. Unless the specimen was small, it was taken back to their workshop in several pieces—what may be called the "brown paper bag" method of collecting. Mary Anning collected this particular skeleton by removing the individual bones from the matrix, because the marl was so soft. This simplified her collecting job considerably, but she lost the relationships between individual parts, making it difficult to reconstruct the skeleton afterward.

Most of the skeletons she collected were taken out of the field in slabs, rather than as single bones. The uncovering of the skeleton in the field would not have been very thorough, so she would then have had to scrape the remaining matrix that adhered to the exposed bone surfaces. Some repairs would likely also be required. Finally, the whole specimen would have to be consolidated. The method used at that time was to assemble the slabs within a wooden frame, in their correct juxtaposition, pouring in plaster of paris to help bind the whole thing together. Sometimes, to reduce costs, a mixture of sand and hot wax was used instead. Once this had set it was finished off with a thin layer of plaster. The resulting specimen, embedded in its wooden frame, could then

William Buckland, during his early years at Oxford University.

be mounted, like a picture, on a wall. Similar methods are still sometimes used today to deal with these kinds of fossils, the end product being called a wall or slab mount.

Mary Anning sold her specimen, probably in several pieces, to Colonel Birch, a fossil collector. Judged from the prices fossil reptiles fetched back then, she probably received between £50 and £100, which would have made a considerable difference to her beleaguered family.

Birch made the remarkable new sea creature available to the Reverend William Conybeare to study. Conybeare named the group to which it belongs "plesiosaur," meaning "near reptile." This is because he thought the creature was closer to modern reptiles than were the ichthyosaurs. We cannot be sure what Anning thought about her new discovery, or about ichthyosaurs for that matter, because she left no record. But, like Buckland and Conybeare, she would have seen them as creatures that lived in the sea in the remote past, long before the time of the great biblical flood of Noah. The flood represented a finite time marker between the remote prehistoric past and historical times—a safety barrier, keeping the revealed world of God-fearing souls safe from the antediluvian creatures and their world of bestial struggle.

There is no record of Anning having had any visitors during the winter she discovered the plesiosaur. But there was one person who regularly visited Lyme at that time of the year—William Buckland. His black mare was a familiar sight, plodding unhurriedly along the shore, her rider slouched comfortably betwixt saddle and top hat. And there was always his famous blue cloth bag slung over one shoulder—he was seldom seen without it. He had a round, almost cherubic face, much prone to laughter, with eyes that always seemed to be smiling, even on drab winter days. "His geniality and love of humour, and even of downright fun, make him a charming companion."

Buckland's trip from Oxford to Lyme Regis, a distance of some 100 miles, must have taken his old mare many days, heavily laden with bags and boxes and sundry geological tools. His journey would have taken him down footpaths and cart tracks, and along some of the most atrocious stretches of roads, muddied and rutted by legions of cartwheels. There would have been many stops along the way at quarries and other sites of geological interest. His favorite mount

> Soon learnt her duty, and seemed to take an interest in her master's pursuits; for she would remain quiet, without any one to hold her while he was examining sections and strata. . . . she became so accustomed to the work that she invariably came to a full stop at a stone quarry, and nothing would persuade her to proceed until the rider had got off. . . .

Buckland, who started teaching at Oxford in 1813, regularly visited Lyme Regis during vacation times. The Christmas break, between the Michaelmas and Lent terms, was a favorite time, and he would often encounter Mary

Anning along the way. She likely would have welcomed the opportunity to chat with him about fossils, because, aside from the sisters Philpot, there were few people in Lyme with whom she could have an intellectual discussion. A stranger happening upon the scene might have been startled at the ease with which the Dorset native conversed with the Oxford professor. Few people of her lowly birth, especially one of her disenfranchised sex, would have dared do more than respectfully pass the time of day with a gentleman, or with a lady for that matter. Part of the reason for her lack of discomfiture stemmed from their long acquaintance, and from his eccentric outlook on life. Part, too, was attributed to their common and consuming interest in fossils. But an increasingly large part had to do with her growing confidence in her own abilities. Buckland had been collecting fossils since before she was born, but, although he was a competent fossilist, he seldom spent more than a week or so in the field at any one time. In contrast, Anning was out collecting fossils almost every day. Most of the fossils in his collection had been purchased from other collectors, including the Annings, but she did all her own collecting. So, in spite of his seniority, she was the more experienced collector.

A greater contrast between two collectors—one for curiosity, the other for necessity—could hardly be imagined. And after they had completed their respective day's fossiling, they would make their separate ways back to their respective accommodations. Both would be cold, tired, and hungry. She would make her way up the beach, through a gap between the buildings where the Marine Theatre stands today, thereby gaining access to Church Street. A short walk would bring her to a pair of cottages, joined by a common wall. The left-hand cottage was the one the Annings rented. Moments later and she would be off the street and standing inside a small and cluttered room, unlit, unheated, and unwelcoming.

If she had continued a little farther along the street she would have walked past the Three Cups, on the opposite side of the road. This was Lyme's best-appointed inn, catering to the most affluent of visitors. And this is where Buckland might have been found at the end of the day, ensconced within a cozy dining room, warming his buttocks before a roaring fire while a haunch of beef sizzled in the grate.

3

The Scriptures and the Rocks

William Buckland was larger than life, and his enthusiasm for geology and paleontology must have been self-evident to all who met him. Sir Roderick Murchison, a contemporary geologist, describes visiting him in his rooms in college:

> On repairing . . . to Buckland's domicile, I never can forget the scene that awaited me. Having . . . climbed up a narrow staircase, I entered a long corridor-like room, which was filled with rocks, shells, and bones in dire confusion, and in a sort of sanctum at the end was my friend in his black gown, looking like a necromancer, sitting on one rickety chair covered with some fossils, and clearing out a fossil bone from the matrix.

Buckland was the consummate actor, and went to great lengths to entertain, as well as to enlighten, his audience. One of his former students recalls the first Buckland lecture he ever attended:

> He paced like a Franciscan Preacher up and down behind a long show-case. . . . He had in his hand a huge hyena's skull. He suddenly dashed down the steps—rushed, skull in hand, at the first undergraduate on the front bench—and shouted, "What rules the world?" The youth, terrified, threw himself against the next back seat, and answered not a word. He rushed then on me, pointing the hyena full in my face—"What rules the world?" "Haven't an idea," I said. "The stomach, sir," he cried (again

mounting his rostrum), "rules the world. The great ones eat the less, and the less the lesser still."

Buckland's eccentric ways, and his zeal for the new sciences of geology and paleontology, must have caused some consternation among the older dons. One venerable faculty member, commenting on Buckland's recent departure on an extended trip to Europe, remarked, "Well, Buckland is gone to Italy; so thank God we shall hear no more of *this geology*." The older faculty members could remember times, not long past, when science had no place in the curriculum of Oxford's undergraduates. Some academics would probably have preferred to keep it that way too. Theology had always played a large part in academic life at Oxford, and most of the staff were ministers of the Anglican Church. Many students, Buckland included, became ordained ministers themselves after graduating, often entering upon ecclesiastical careers, as did his close friend and confidant, William Conybeare. But the university had come under increasing pressure during the early part of the nineteenth century to break out of its academic straitjacket.

Oxford's scholastic high table offered rich fare for her privileged browsers, all of whom were men. The main courses comprised history and theology, with generous helpings of philosophy and the classics, and side dishes of mathematics and geography. Students could also attend classes in chemistry and anatomy, though anyone aspiring to a medical qualification had to complete his studies elsewhere, primarily at London University or Edinburgh University.

By 1809 Professor John Kidd was offering a course of lectures in mineralogy, which he broadened to include the new and rapidly growing science of geology. Kidd's lectures whetted Buckland's appetite for geology, but others found the subject totally indigestible. The poet Percy Bysshe Shelley went off to his first lecture with great enthusiasm, but he left before it was finished, confiding in a fellow student that "it was so stupid and I was so cold that my teeth chattered." When asked what the professor had talked about, Shelley replied: "About stones!—stones, stones, stones! nothing but stones!—and so drily."

Kidd subsequently accepted a Readership in Anatomy, and a successor was sought for his mineralogical duties. Buckland and Conybeare were both considered for the position of Reader in Mineralogy, a position equivalent to a North American full professor. Kidd's first choice was Conybeare, whose knowledge of geology was far greater than Buckland's. But Conybeare's loyalty to the church and his financial independence (he inherited a small annuity from his grandmother) led him to decline the position and become a minister of the church, though he continued to be an influential force in geological and paleontological circles for many years to come. Buckland therefore attained the position through default, in 1813.

Buckland was a great believer in the importance of visual aids in his teaching. He drew large wall charts and amassed a significant collection of geological and paleontological specimens. Among the several geological charts and maps he constructed was one showing the succession of the strata, as far as was known, for the British Isles.

The pioneering work of mapping the various rock layers across the land had been undertaken by William Smith (1769–1839). Smith, a surveyor, had noticed that different rock layers could be recognized by the characteristic fossils they contained. And these layers, or strata, could be recognized over large distances. For example, certain species of sea urchins that occur in the chalk of the southern counties of Sussex and Kent can also be seen farther north in the chalk of Cambridgeshire and Yorkshire. But Smith's maps were not published until 1815, so Buckland had to cobble together his own.

By means of his charts, maps, and fossils, Buckland could show his students the orderly succession of events through time, as written in the rocks. There was a time, long ago, before the Earth was inhabited by living things. This inanimate period was evidenced by the primitive rocks, which contained no fossils. The early geologists later referred to these as primary rocks—what we today refer to as igneous rocks. Buckland taught that these primitive rocks, as exemplified by granites, predominated in the Earth's crust, and occupied the lowermost position in the stratigraphic sequence. Buckland judged them to have reached a total thickness of at least two miles. Lying above the primitive rocks were the secondary rocks. These formed slowly, over long periods of time, by the accumulation of sediments—hence the modern term sedimentary rocks—and contained the fossilized remains of animals and plants. He used his fossil collection to illustrate the orderly succession of life through time. One of the large ammonites from the foreshore of Lyme Regis might be used as an example of a creature from the lower strata, while a mammoth tusk might illustrate a creature from the superficial deposits. The mammoth's superior position in the stratigraphic sequence showed it was among the more recent products of God's creative wisdom. Although such creatures no longer existed, they lived in Britain during the far-off antediluvian period.

Geology, like most other branches of science, is a practical subject, and Buckland spent a good deal of his time in the field. His geological excursions,

Buckland's entertaining lectures were well attended by students and faculty.

often in the company of his close friend Conybeare, took him to all parts of the country, and to many parts of the Continent. He was a firm believer in teaching geology in situ, and organized frequent field trips for his students. He would often make an announcement at the end of his lecture to the effect that tomorrow's class would meet at the top of Shotover Hill, or at the quarries of Stonesfield. Both localities were good for fossil hunting, and were only a short horseback ride from Oxford. But he sometimes took them much farther afield, usually giving them an impromptu lecture when they got there. He was an inveterate prankster, often playing tricks on his class when the opportunity arose. One of his favorite ruses was to wait for a rainy day, then get the novice students mired in glutinous marl—he thought it important to show them how treacherous such terrain could be for the unwary.

Buckland's lectures were always well attended, both by students and members of the teaching staff. Many came to be enlightened, others came to be amused and entertained. But most attended to be kept abreast of developments in the exciting new sciences of geology and paleontology, which were as young and vibrant then as genetic engineering is today. Buckland, still in his early thirties, was at the cutting edge of his field.

The stipend Buckland received for the Readership in Mineralogy was £100 a year, provided by an endowment from the Prince Regent. This was a rather poor income, even for those times, being little better than that of a porter. However, his accommodation, and presumably his meals, were provided, and he was still a bachelor, with few expenses other than books, fossils, and traveling. Writing of his circumstances to his uncle in 1818 he says he was content, but that, "I crave to be lecturer in Geology—for that I ask £100 for my lectures; and £100 for being curator of the collections . . . "

He compared his income with that of the Professor of Mineralogy of Dublin, who received three times as much. In an attempt to improve his lot, Buckland wrote a memorial to His Royal Highness the Prince Regent, for additional support. Choosing his words carefully, Buckland made a strong case for geology, which he described as being, "so much cultivated, of so much National importance, and so liable to be perverted . . . [against] the interests of Revealed Religion. . . . "

The Prince Regent was one of the greatest patrons of the arts to sit on the English throne, and Buckland's reference to culture was likely to appeal to his readiness to support things considered to be in good taste. Buckland's claim of national importance might have struck a chord with the prince too, because geology played a vital role in securing raw materials to fuel the factories and smelters of the Industrial Revolution. However, Buckland's concern that geology should not be used to undermine religion might have fallen on deaf ears: The prince had the morals of an alley cat, and his licentious behavior during the stewardship of his father's throne—a duty necessitated by King George III's madness—brought the nation's scorn upon the monarchy. But the prince may have been concerned for the political stability the church engendered. These were politically tumultuous times, both in England and across the Channel. Just two decades before, France had been torn apart by a revolution that had overthrown the monarchy, separating royal heads from royal shoulders. The Prince Regent had a vested interest in protecting the status quo. Buckland got his increment, and also received the new title of Reader of Geology.

As Reader of Geology, Buckland was required to give an inaugural lecture. His passionate desire to reconcile geology and the Scriptures led him to make this issue the subject of his lecture.

For all his panache at the podium, Buckland was always uneasy when confronting a new audience. "[H]e feels very nervous in addressing large assemblies," a contemporary wrote, "till he has once made them laugh, and then he is entirely at his ease." But there would be no laughter during his inaugural address: Lecturing in that hallowed hall of learning was a stiffly formal affair, even during regular classes. The young gentlemen, in their academic gowns, took up the back and middle seats, the front row being occupied by the academic staff. We can appreciate his apprehension as he gazed out over the sea of expectant faces on that fateful May day in 1819, the front row swollen to capacity with the highest ranking officers of the university. At the appointed moment he stood up, took a deep breath, and launched into his lecture.

When he had finished his audience showed their approval with a spirited round of applause. They had just witnessed Buckland's views on how geology, far from refuting the Scriptures as some people feared, completely supported the biblical account of the early history of the world. At least, they *thought* they had heard Buckland's views on the subject, but most of the ideas he expressed were attributable to others, mostly to Conybeare, and the lecture had little original content.

At the time of his inaugural address, Buckland had been teaching geology for five years. He was therefore well versed in his subject, and in the art of presenting information to audiences. He was a pragmatist and empiricist, a mud-on-his-boots field man, but when it came to more philosophical issues he drew heavily upon others, primarily upon Conybeare. Reconciling the Bible with the geological record was an exceedingly complex issue, and, in anticipation of the great task that lay ahead of him, Buckland had written to Conybeare several months before the inaugural lecture. Conybeare responded with a huge missive of twelve folio pages, filled with his neatly sloping handwriting.

> "I am much delighted with your letter," he began. "It gives me sincere pleasure to see your career of science becoming daily more brilliant. I highly approve the sketch you have given of topics for an inaugural lecture. Make it a classical composition and publish it."

Conybeare gave detailed notes for the lecture, answering questions Buckland had raised and giving references to further reading. The tone of the letter was very much that of a teacher addressing his student: ". . . you must say something of the progress of Geology in Oxford, and you should . . . pay a few compliments to Kidd. . . . "

Conybeare's letter ended with a postscript: "You can afford to pay postage with your fat salary." (Until the first postage stamp, the Penny Black, was introduced in 1840, it was the recipient who paid for the cost of mail delivery.)

William Conybeare, probably the brightest member of the geological circle during his day.

Buckland drew freely from Conybeare's letter for his lecture, copying parts of it verbatim and expanding on his ideas and suggestions. He also quoted extensively from other sources. As a consequence, less than one-third of his entire lecture was original. That is not to detract from its importance or to suggest that he was not in complete accord with Conybeare's philosophy. The point is that Buckland, who was not such an original thinker as Conybeare, was still marshaling his own thoughts on how the geological evidence could be used to support the biblical account of the Creation, which his religious beliefs told him was true.

In his preamble to the lecture Buckland justified the importance of the new science of geology and its relationships to other branches of science. He emphasized the incompleteness of the fossil record, a point Darwin would belabor some forty years later in his *Origin of Species.*

The fossil record was indeed lamentably incomplete in 1819. There were no dinosaurs or plesiosaurs, the ichthyosaurs had not long been discovered, and the flying reptiles, the pterosaurs, were barely known. Even today, after all the thousands of fossil species that have been discovered, we still have only a small sample of the many millions of species that formerly inhabited the planet.

Buckland started on the main point of his lecture—the geological corroboration of religious truth—by noting how the most valuable mineral resources, like coal and metallic ores, were the most abundant. They were also located where they could be most conveniently extracted. The coal layers, for example, were inclined at an angle, rather than being horizontal in position, and this facilitated mining operations. Furthermore, the occurrence of breaks in the sequences—faults—reduced both the risks of flooding and the spread of fires. Similarly, hills and valleys were "curiously constructed" to catch occasional rains, thereby providing an everlasting supply of freshwater, through rivers and springs. "In all of these we find such undeniable proofs . . . of wise foresight and benevolent intention and infinite power, that he must be blind indeed, who refuses to recognize . . . proofs of the attributes of the Creator."

Buckland saw the same "proof" of the wise Creator in the design of living organisms, each one forming a "link of the boundless chain of living beings . . ." Quoting freely from Archdeacon William Paley's *Natural Theology* (1802), every example of natural design, from the movements of the planets to the interlocking mechanism of the barbs of a bird's feather, were cited as incontrovertible "proofs" of the existence of the "supreme intelligent Author. . . . "

The supposed "proof" of the existence of God on the grounds of design was a popular and persuasive argument throughout much of the nineteenth century. The same banal argument still finds favor today among certain fundamentalists posing as scientists—the proponents of "creation science."

Buckland gave two examples, "two great points" of concordance between geology and the Bible: "the low antiquity of the human race" and "the grand fact of an universal deluge. . . . " The recency of our species was clear because human remains were found in only the most superficial deposits, and never in association with extinct animals. The geological evidence for the Noachian flood was provided for by the way river valleys have been carved by the floodwaters and by the immense deposits of gravel left behind by the receding waters. Quoting from Cuvier, Buckland placed the age of the Noachian flood as not more than 6,000 years ago. But this did not place any limitations on the antiquity of the antediluvian Earth. Nor did it preclude the existence of ancient life-forms that predated the creation of humans and contemporaneous creatures: "though Moses confines the detail of his history to the preparation of this globe for the reception of the human race," Buckland told his attentive audience, "he does not deny the prior existence of another system of things . . . "

Buckland concluded his lecture by suggesting that the discrepancy in time scales between Genesis and geology was best explained by assuming that "the word 'beginning' as applied by Moses in the first verse of the Book of Genesis, [was used] to express an undefined period of time which was antecedent to the last great change that affected the surface of the earth [the Noachian flood]." A succession of creations appeared and disappeared during this long time period, as the world was swept by one revolution after another, culminating in the appearance of modern animals and humans.

Buckland's views of Earth history can be summarized: The Earth was created by God a considerably long time ago, perhaps hundreds of thousands of years. At first the Earth was devoid of life, and this period was represented by the great thickness of "primary" rocks that were devoid of fossils. As told in the Book of Genesis, God created different creatures at different times, humans being the last. This was evidenced by the succession of different kinds of fossils in the geological record, with the more primitive creatures, like reptiles, being found in the lower strata, mammals higher up, and humans in the most superficial strata.

His inaugural lecture was a great triumph, vindicating his "favorite science of geology . . . of the charges that have been brought against it. . . . " He could therefore go about his fossiling and geologizing with the blessing of the church, and the approval of most, if not all, of his clerical colleagues at Oxford.

However implausible, and however ridiculous Buckland's "proofs" of the existence of God may appear to us today, we have to put things into the correct historical context—we must put ourselves in his shoes. Buckland and his contemporaries did not recognize the distinction between natural phenomena, which could be investigated scientifically, and religious convictions, which lay beyond the purview of science. Furthermore, although they often used the term "science," their understanding of the word was not the same as ours. The early fossilists were simply not scientists—a term that only came into usage later in their century—and had therefore not been trained in the scientific method. In science, inferences are drawn from observations. These inferences form the basis of hypotheses, which can be tested by appropriate experiments and further observations. For example, a biologist might infer from the observations of skeletons in museums that large animals have more robust leg bones than smaller ones. A reasonable hypothesis from this inference is that large animals need relatively thicker limb bones to bear the larger loads on their heavier bodies. This (incorrect) hypothesis could be tested by plotting a graph of limb robustness against body weight, to see if there were a correlation.

Buckland was being totally unscientific when he inferred from his (superficial) observations on the distribution of minerals that they had been so placed by God for the benefit of humans. After all, what possible experiment or further observation could he have devised to test the inferences of the intent of God? Such things are clearly untestable and lie outside the bounds of science.

But the early fossilists were not always unscientific in their interpretations of the natural world. Cuvier, for example, was being perfectly scientific when he inferred that the sharp teeth and sharp claws of fossil animals, like living ones, were correlated with carnivory. And Buckland, for all his illogical lapses, was quite capable of approaching a problem scientifically. For example, there were stories abroad during his time that toads were sometimes found entombed alive inside rocks. Buckland did not think toads could survive such conditions, and conducted some experiments where he buried live toads in stone containers. Bottles of champagne were wagered on the outcome of his experiments at one of the meetings of the Geological Society. Predictably, all the toads died, vindicating his position.

Buckland is remembered for many things, but his most enduring legacy to paleontology concerns a discovery he made not far from his home. Eight miles northwest of Oxford, nestled among gently rolling hills, lies the small village of Stonesfield. It is a picture-perfect place of stone cottages and dry stone walls, and during Buckland's time the surrounding fields were dotted with stone quarries. Stonesfield was a prime fossil-collecting locality, and Buckland frequently visited the quarries, often accompanied by his students or by Conybeare. The rocks are Jurassic in age, dating back some 150 million years. At that time, this part of England, which lay next to the sea, had abundant lakes and rivers. This was revealed by the discovery of marine and freshwater fossils, mingled with those of terrestrial origin. Buckland's potpourri of Stonesfield fossils included sharks' teeth and fish scales, crocodile bones and teeth, tortoiseshell, ferns, and wood. Buckland also discovered some exceptionally large bones—twice the size of our own. He probably thought they belonged to

Georges Cuvier, one of the greatest intellects of his time, was the world authority on anatomy.

some kind of giant crocodile, and added them to his large and growing collection without giving them much more consideration: Buckland was not always the most meticulous field man, as his former student, Charles Lyell, once confided to Mantell.

In the year before Buckland's inaugural lecture, Cuvier had written to George Greenough (1778–1855), one of the founding members of the Geological Society, asking him to make arrangements for his forthcoming trip to England. Cuvier wished to visit the Royal College of Surgeons' Hunterian Museum, which housed an enormous collection of skeletons and pickled specimens, collected by John Hunter, the celebrated surgeon, who died in 1793. Cuvier also wanted to meet Buckland. This was natural enough—Buckland was the preeminent British fossilist at the time. Arrangements were duly made, and the two met in London a week later. After their London meeting Cuvier accompanied Buckland to Oxford to see his fossil collection. Cuvier's English was minimal, and it is likely they conversed in French. There were probably many items in Buckland's collection that interested the great French anatomist, but the ones that riveted his attention were the giant bones from Stonesfield. It is unclear how much of this material Buckland had at the time, but he certainly had a femur, and probably at least one vertebra. Cuvier had seen bones like these before, collected from the sea cliffs of Honfleur in Normandy. Cuvier thought they were the remains of some sort of crocodile, but a kind totally unlike any living ones. The Honfleur crocodiles were huge too. Cuvier gave a length estimate of over forty feet, which must have impressed Buckland.

Cuvier returned home to France, convinced that giant reptiles once roamed the Earth—reptiles that were considerably larger than their living counterparts. His visit probably piqued Buckland's interest in his Stonesfield discovery, but it would take him another five years before he did much more about it. Why did Buckland not act sooner? The observations of Cuvier's artist and loyal friend, Charles Laurillard, may be insightful here. Writing home to France during a London trip with Cuvier the previous year, Laurillard commented:

> We were able to see very beautiful things in natural history and mainly in fossils. Their collections cannot compare with ours. The College of Surgeons holds very nice preparations done by Hunter, and also some skeletons, but neither the English anatomists, nor the naturalists, are too well versed in zoology or comparative anatomy so they do not even know the value of their riches.

Perhaps not, but that did not stop English fossilists from looking. And Buckland was not the only one finding the remains of giant reptiles on English soil. Seventy miles to the south, in the picturesque county of Sussex, Gideon Mantell was about to make some startling discoveries of his own.

4

Chalk Pits and Leeches

Cuckfield, a small Sussex village a dozen miles north of Lewes, was the first refreshment stop on the stagecoach route from the fashionable seaside town of Brighton to London. Brighton's brisk growth had resulted in a rapid expansion of coach traffic to the capital, and this, in turn, led to an increased demand for gravel and stone for road maintenance. Much of this need was met by a large quarry at Whitemans Green, within sight of Cuckfield. The workmen smashed the rocks they quarried into pebble-size pieces, and in the process they occasionally found fossils. These were usually rather badly damaged by this stage, except the smallest ones, so the fossils from Cuckfield were largely fragmentary. But to Gideon Algernon Mantell, doctor, geologist, museum curator, author, and passionate fossil collector, the quarry was a gold mine. What made these fossils so special was that they seemed to represent an entirely different fauna from his other Sussex fossils. They were also geologically older.

Mantell lived in Lewes, a small and unspoiled county town, nestled among the chalk downs of Sussex, one of the most charming of the English counties. Sussex is a land of rolling hills and beech woods, meadows and golden fields, a pastoral counterpane thrown over an undulating bed of sedimentary rocks. Almost all these rocks belong to the Cretaceous Period, which immediately follows the Jurassic. Lazy rivers meander to the sea, and the edge of the land is so thick in parts that terra firma often ends abruptly in sheer cliffs of gleaming white chalk. The chalk was formed on an ancient seabed, more than 65 million years ago, by the accumulation of the calcareous skeletons of dead microorganisms raining down from above. The fossils found in chalk are almost exclusively marine, with only occasional terrestrial forms carried into the sea from the nearby land. The fossils are often abundant and well preserved.

Mantell had acquired most of his fossils from the chalk, and had a large collection, mostly of invertebrates. There were corals and sea urchins, starfish and sea lilies (crinoids), ammonites of all shapes and sizes, and seashells of every description, including lampshells, or brachiopods. There were also some vertebrate fossils, including sharks' teeth and fish scales, and parts of the skeletons of fish.

The chalk is part of the last stage of the Cretaceous, referred to as the Upper Cretaceous, and it marks the end of the Mesozoic Era, which is often called the Age of Reptiles. Almost all the rest of the Sussex strata are much older, belonging to the Lower Cretaceous. Today these strata are referred to as the Wealden, a name derived from the Old English word for woods, which once covered most of southern England. The Cuckfield quarry was Wealden in age.

According to the best knowledge available at the time, the fossils from the Lower Cretaceous, like those from the chalk, were all marine. But when Mantell examined the Cuckfield fossils, he was surprised to find none of the typical marine animals that were so abundant in the chalk. There were no ammonites, no corals, and no brachiopods. Furthermore, there were some remains of land plants, together with fragments of animals, like amphibians, which were known to live in freshwater. There were some marine fossils too, but the presence of the terrestrial and freshwater ones showed that Cuckfield was once close to the shore. He called the new strata, with its mixture of terrestrial, freshwater, and marine fossils, the "Strata of Tilgate Forest," named after an historical wooded area.

Finding terrestrial animals in rocks of Cretaceous age was a novel prospect at that time, so Mantell wanted to be sure of his facts. This required more fossils. To this end he paid more visits to Cuckfield and to other localities where he might collect Tilgate fossils. Mantell recalled how, "many a long and weary journey have I undertaken, to examine the materials thrown up from a newly-made well, or . . . cuttings on the road-side, in the hope of obtaining data by which the problem might be solved . . . ." Mantell began acquiring fossils from Cuckfield at least as early as 1819. There is an entry in his journal for June 30 recording that "A packet of fossils from Cuckfield from Mr. Leney arrived, it contained bones, teeth. . . . " He also found Tilgate fossils at some other quarries.

In the fall of 1821 Mantell was paid a visit by Lyell. Although they had never met before, they sat talking about geology until late into the night. The two men, who became firm friends and confidants, continued their conversation the following day. Something of considerable importance had clearly seized their attention. The letters they exchanged afterward reveal that the main topic of their discussion was the fossils from Cuckfield. By this time Mantell had probably accumulated enough Tilgate Forest material to be fairly confident he had discovered a new fossil assemblage. Here was a tantalizing glimpse of some of the plants and animals that lived on land during the remote Cretaceous period. This discovery alone would have caused Lyell's geological pulse to race, but this was not all that Mantell had discovered. He had also unearthed the bones of some gigantic animals, fragments of which began appearing at Cuckfield at least as early as the summer of 1820.

We can only imagine what must have been going through Mantell's mind as he wrestled with the problem of his Tilgate discoveries. He was a relative newcomer to geology, working essentially in isolation, too busy with his medical practice and too isolated from London to have ready access to the other members of the geological circle. He had still not met Buckland, the acknowledged authority on fossils. In spite of his isolation, he had discovered evidence that was contrary to contemporary understanding. The Cretaceous rocks should have been exclusively marine, but they were not. As if that were not enough, he had discovered the fragmentary remains of some gigantic animal, quite unlike anything he had ever seen before. Even the quarrymen, well used to finding all manner of odd things in the rocks, had never seen fossils like these. He probably recognized from the start that they were the bones of some colossal terrestrial animal rather than some sort of whale. But what sort of animal was it? Could the elephant-sized bones belong to some sort of extinct pachyderm?

Lyell probably saw the giant bones during his first visit to Lewes. He would have told Mantell of the equally colossal bones Buckland had been collecting from Stonesfield. Indeed, there was a remarkable similarity between the two assemblages, each showing a mixture of terrestrial, freshwater, and marine fossils. No doubt they discussed this similarity at great length. When Lyell returned to London he sent Mantell a small sample of Stonesfield fossils, for comparison with his Tilgate assemblage:

> Unfortunately neither Buckland nor any of the best geologists were at Oxford the week I spent there," Lyell wrote to Mantell, after sending his first package of fossils to him, "so I could not see Buckland's fine collection. . . . had a hasty visit to Stonesfield where I procured a good box full, many the same as what I sent you. . . .

Lyell went on to ask, "What weight of evidence do you require to identify beds?," wondering aloud whether the Stonesfield and Tilgate fossil assemblages were the same. "You say you detect decided differences in many of their organic remains," he continued, but, "How is it possible . . . both contain amphibia, aves, pisces, insecta, vegetables mixed together . . . ."

Lyell went to a great deal of trouble on behalf of his new friend: "I have examined Greenough's collection and compile you a list of his Stonesfield fossils. . . . " His list included a large bone, probably similar to the ones Buckland had shown Cuvier. He also included an extract of a letter Buckland had written to his friend George Greenough, saying, "Cuvier has no doubt that the great Stonesfield beast was a monitor [lizard] 40 foot long and as big as an elephant."

Lyell was completing his law studies at the time, and concluded his letter with a plea to Mantell not to write back before Christmas, "for I am buried in the study of law [and] I am too fond of geology to do both. . . . "

Mantell understood perfectly. Although seriously committed to medicine and to his patients, his passion in life was for fossils and geology. But he felt rather isolated from what was going on in the field, ensconced in his bucolic backwater. He had joined the Geological Society the previous year, but the six-hour coach journey to London and the pressures of work made it difficult for him to attend meetings on a regular basis. He probably hoped his new discoveries would carve a niche for him among his newfound circle of geological colleagues.

Regardless of the obvious similarities between the Stonesfield and Tilgate assemblages, Mantell and Buckland eventually recognized that they were not contemporaneous and were not part of the same geological formation. The Stonesfield strata lay in the middle of the Jurassic sequence, while the Tilgate strata were at the bottom of the Cretaceous. This raised the strong possibility that the giant bone fragments Mantell had found in Sussex did not belong to the same beast as the large bones Buckland had discovered in Oxfordshire. That two entirely different kinds of unknown giants had stalked the Earth during the remote past was an exciting prospect, and raised the possibility that there might be other bizarre creatures waiting to be discovered.

As 1821 drew to a close, Mantell's first book, *The Fossils of the South Downs; or Illustrations of the Geology of Sussex*, was nearing completion. Although he had a publisher, he had to pay the entire production costs himself. He met this expense by selling subscriptions to potential purchasers. His entreaty to the palace won him the patronage of King George IV (formerly the Prince

Gideon Algernon Mantell and his wife, Mary Ann Mantell.

Regent), who commanded that his name be placed at the top of the list for four copies.

Mantell's subscription list increased more rapidly than he had dared hope for. This was a reflection of the growing public interest in the new sciences of geology and paleontology. It was also a portent of Mantell's future role as a popularizer of science.

Mantell had been working on the book for almost four years and must have been feeling relieved as the end drew in sight. Most of the illustrations had already been printed, and he had almost completed the writing. Or so he thought. But then something quite remarkable happened, something that would require him to write a whole new section. We might expect to find a full account of this singular development in his journal, which recorded such details as the price of medicinal leeches (£1 for 200) and the sentences passed on thieves (usually hanging). But there was not a single mention of the most significant paleontological discovery of his life.

According to the oft repeated story, Mrs. Mantell, the dutiful wife, was waiting beside the carriage while her husband made a professional call. As she whiled away the time something by the roadside caught her eye. It was something small, embedded in a piece of rock that was part of the road material.

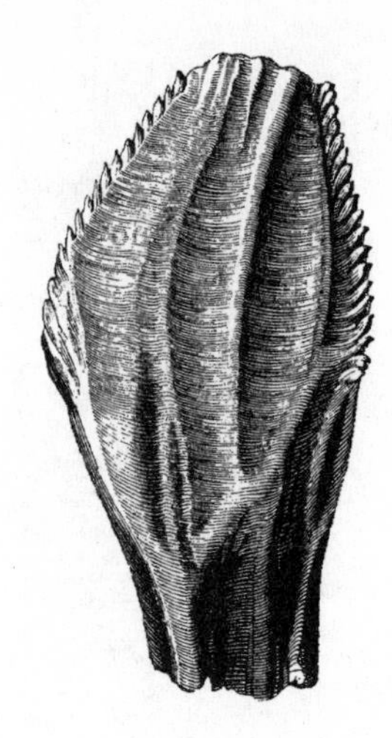

One of the remarkable teeth that Mantell discovered.

When she showed the walnut-sized object to her husband he was ecstatic—she had discovered the tooth of a very large and unknown animal. The source of the rock was subsequently traced back to the Cuckfield quarry, where more teeth, and some bones, were eventually unearthed.

This is a wonderful story of the discovery of one of the most important fossils of the time. Unfortunately, it is without foundation, and Mantell himself was largely to blame for this. Not only did he fail to record at the time how the unique tooth was discovered, but he also gave somewhat different accounts of the find after the fact. The events leading up to the discovery will therefore always be clouded in doubt. Dennis Dean, an authority on Mantell, is of the opinion that the first tooth was probably supplied to Mantell by Mr. Leney, the quarryman at Cuckfield. He also suspects that Mantell was in possession of several teeth by November 1821. Regardless of the provenance of the first tooth, it must have astonished and baffled Mantell because it was so entirely different from anything he had ever seen before.

The teeth come in a variety of shapes and sizes but they are commonly about two inches long (5 centimeters), and as thick as a thumb. Like the terminal joint of a thumb, the tooth is somewhat flattened. One surface, the "ball of the thumb," is rounded. The other surface is fluted by three broad grooves. The upper half or so of the tooth, the crown, has a serrated margin—it feels like a coarse file. The rest of the tooth, the root, gently tapers to a blunt termination. A flat wear-facet can sometimes be seen at the tip of the crown, and this shows where the tooth has been abraded during use. Wear facets like this are typical of herbivores because plant material is very abrasive, a fact that was not missed by Mantell, who commented that the tooth: "so entirely resembled . . . an incisor of a large pachyderm, ground down by use, that I was much embarrassed to account for its presence in such ancient strata; in which, according to all geological experience, no fossil remains of mammalia would ever be discovered. . . . "

He was therefore reluctant to identify the tooth as being mammalian because, at that time, mammals were not known to exist in rocks as old as the Cretaceous. We now know that mammals did live during the Mesozoic Era alongside dinosaurs, but they were diminutive, and did not reach the large sizes of pachyderms. He correctly concluded that the unusual teeth belonged to a land animal, and, since mammals were excluded, this seemed to leave reptiles as the only alternative. However, Mantell knew that "no known existing reptiles are capable of masticating their food," so he "could not venture to assign the tooth in question to a saurian." But what if the tooth belonged to a reptile completely unlike any modern kind? Mantell was leaning toward this idea, and probably suspected that the massive bone fragments he had discovered from the same locality belonged to the same kind of creature too.

He took the tooth along to a meeting of the Geological Society during the summer of 1822 and showed it to several learned members, including William Clift (1775–1849), a respected anatomist who worked at the Royal College of Physicians' Hunterian Museum. By now Mantell was convinced he had discovered an entirely new kind of animal—probably a giant herbivorous reptile—and probably anticipated the enthusiastic endorsement of the other gentlemen. He was therefore bitterly disappointed when they dismissed the tooth as being of no particular consequence. They suggested it was probably a fish tooth, or perhaps a mammal tooth, displaced from a more recent rock layer. Despite their discouraging words Mantell remained convinced he had discovered a new kind of reptile. But he lacked the evidence and the confidence to commit these thoughts to paper. He therefore wrote a cautious additional section to his book, giving a brief description of the unusual teeth. But he did not attempt to identify them, nor give any illustrations.

It was sometime during this period that Mantell received a long letter from a theologian, reconciling geology with the Scriptures using some of the arguments Buckland had used in his inaugural lecture. Mantell was sufficiently impressed with the letter to obtain the author's permission to reproduce it as a preliminary essay to his book. In this way he neatly side-stepped any religious conflict his book might pose, allowing him to devote his entire energies to the geological issues at hand, avoiding all the conciliatory discussions that occupied so much of Buckland's time.

Mantell wrote the last page of his book on May Day, 1822. It was published just four months later, and was very well received, both by his geological peers and by lay readers. Mantell was not a man to rest on his laurels, however, and was probably already thinking about his next book. He was also anxiously trying to acquire more material of the animal with the enigmatic teeth, to establish its identity. His network of quarrymen were the most likely source of new material. His cultivation of their loyalty over the years had assured him of first refusal of anything they unearthed. But he was in for a rude shock when he visited one particular local collector, namely the Reverend H. Harper: "Mr. Harper had several splendid specimens of fossils which he had obtained from my workmen at Lewes chalk-pits: this kind of poaching has become so general since my work [the book] has appeared, that I have now no hope of adding anything interesting to my collection."

Nevertheless, he continued to collect Tilgate fossils, hoping to obtain more definitive clues to the identity of his unknown giant. Meanwhile, his friend Lyell, who was about to take an extended trip to Paris in June 1823, offered to take one of the enigmatic teeth to show Cuvier, the ultimate authority on extinct animals.

Mantell anxiously awaited news from Paris, but his busy medical practice kept his mind off matters. As a physician in a country practice he did everything, from treating diseases and broken limbs to performing cataract surgery and excising tumors. But his specialty was obstetrics. And at a time when fourteen women died for every 1,000 births, Mantell lost only two patients in 2,400 deliveries—an outstanding achievement.

Accompanying him on his rounds would have been a horrifying experience for us today. There were no anaesthetics. Consequently, patients who required surgery faced the most agonizing ordeal of their lives ameliorated only by laudanum (opium dissolved in alcohol) or alcohol. There were no antiseptics either, so the chances of surgical incisions or wounds becoming infected were very high. And because there were no antibiotics, these infections could not be treated. Even minor wounds could be fatal. One of Mantell's former assistants pricked his finger during a dissection in that summer of 1823. He died from the infection two weeks later. There was little recognition of the need for cleanliness, and a major reason for the high mortality among women in labor was the spread of infection, notably puerperal fever. The attending physicians were the main culprits, and although some, like Mantell, washed their hands between patients, others did not. (There is one chilling account of a physician who, having performed an autopsy on a woman who died of puerperal fever, immediately attended another labor. He did so without changing his clothes or removing parts of the deceased woman's organs from his coat pocket.) One of the reasons for Mantell's remarkable obstetrics record was that he administered ergot, a powerful constrictor of blood vessels, to deal with the hemorrhaging that sometimes follows childbirth. But, like others of his time, he was ignorant of menstruation, conception, fetal growth, or the mechanism of birth. The standard remedies for most maladies, from sprains and stomach pains to epilepsy and pneumonia, was to use leeches or strong purgatives.

His large and growing medical practice required him to cover many miles of the countryside to visit his patients, usually on horseback, and he often stopped off at chalk pits and quarries along the way, to look for fossils. He spent much time cultivating the local quarrymen to save any fossils they found, so he could purchase them on his return visits. His journal is replete with such entries as: "An enormous ammonite was found in Jenner's Quarry, Malling Hill: sent my servant for this in the evening." Two days later he had "cleared the large ammonite . . . it measured nearly six feet in circumference."

Just how Mantell managed to find time for his fossiling is difficult to imagine, given the frantic pace of his medical practice. In addition to his regular patients and the emergencies he was called upon to attend, he was the medical officer for the parish, and had an appointment with the Royal Artillery Hospital at Ringmer, just outside Lewes. He often visited forty or more patients a day during busy periods, and a staggering sixty-four during the height of a typhus epidemic.

Medicine in those days was in a primitive state of quackery, and physicians counted for little in society, which is probably one of the reasons he strove so hard to seek recognition and status. He tried, and partially succeeded, to seek patients among the gentry, making a conscious effort to move in higher circles. But it was through geology that he hoped to rise above his humble beginnings and his somewhat disreputable profession. His eight geological books and his major geological discoveries eventually won him international recognition, but he was still dissatisfied. He drove himself relentlessly, sleeping only four or five hours each night.

Mantell had to wait several months for news from Paris. He had expected to receive word from Cuvier in the mail, but no letter came. Then his friend returned, the bearer of bad news: Cuvier had dismissed the tooth as the incisor of a rhinoceros. Mantell was crestfallen. He had been so sure he had discovered some new kind of creature, but Cuvier, the highest authority in the world, had relegated the extraordinary teeth to the ordinary.

Mantell could not know it at the time, but the enigmatic teeth were neither mammalian nor unimportant. Indeed, they would eventually be recognized, along with Buckland's finds at Stonesfield, as one of the most important paleontological discoveries of the age. Buckland and Mantell were on the brink of lifting the dark curtain that cloaked the remote past. But at that time Buckland was preoccupied with more recent history: the Noachian flood.

5

The World of Darkness

The Yorkshire moors in winter can be one of the coldest and most desolate spots in England. Winds can howl, driving rain in sheets or snow in blinding flurries. Or the winds can drop, granting silent tenure to the dank mist that stalks the land.

However cold it might have been on the December day in 1821 when Buckland arrived on the moor, he must have noticed how warm it was inside the cave. It would have felt as warm as the European caves had felt cold, relative to the summer's warmth during his Continental tour. This is because cave temperatures stay much the same, summer and winter, providing attractive accommodations for a variety of animals.

Buckland had to crawl on his knees, with a handheld lantern to light the way. The lamp's warm glow was soon lost to the abyssal darkness, and the cold limestone floor would have been brutally hard on his knees. But standing upright, as he had done in the caves in Franconia, was impossible—Kirkdale Cave offers precious few places to stand erect. The passageway is so low and narrow for much of the way that, even while crawling on all fours, explorers have to remember to keep their heads low to avoid painful collisions with the roof. Fortunately for Buckland, the floor was not running with water. Although his visit was at the end of one of the wettest periods in memory, the cave was dry.

The lack of running water at Kirkdale is the reason there are no impressive growths of stalactites from the roof or stalagmites from the floor. Nor are there many flowstones, the encrustations of limestone that grow on the walls of caves, like fungi. Stalactites, stalagmites, and flowstones are formed when water, containing dissolved limestone, flows over the surfaces of a cave, thereby depositing minute particles of limestone. Over the course of thousands of years the drips from the roof grow into frozen daggers of lime, and similar limestone monuments grow from water dripping onto the floor. Water trickling down the cave walls similarly accumulates as flowstones.

Although it has been stripped clean of the fossils that lured Buckland there nearly two centuries ago, Kirkdale Cave is still accessible for exploration, and there is still much to see. The cave does not teem with life, but its sepulchred walls offer sanctuary to some natural curiosities that are seldom seen elsewhere. Gossamer purses, no longer than a small finger, hang from the roof like so many buds. Each one contains the folded body of a hibernating moth. Cave spiders—one of the largest species in Britain—lurk in the darkness, waiting for hapless insects to come their way. Tiny horseshoe bats cling in crevices in their winter slumbers, oblivious to the invasion of their subterranean world. Buckland probably marveled at these wonders too, but his primary objective was to search for fossils.

Word reached Buckland a month earlier that bones had been found in the cave. The discovery had been made the previous summer when workmen accidentally opened the cave mouth, which had become sealed with rubbish and overgrown with bushes. When the workmen found the bones inside they thought they belonged to cattle, and threw them onto the road. Fortunately, they were noticed by a local physician who recognized their importance.

Buckland included a delightful drawing of the entrance to Kirkdale Cave in his first book, *Reliquiae Diluvianae* (Relics of the Deluge). The illustration shows the cave, along with some workmen, and a shovel for scale. The square entrance to the cave is about as high as the shovel and only a few feet above ground level, giving easy access. Not so today. Although the quarry has long since been abandoned, the quarrying activity since Buckland's time has cut the working face back some fifteen feet. Because of the tilt of the cave, its entrance now lies about twenty feet above the ground.

Buckland's drawings include an informative cross-section and a diagrammatic plan view that shows the cave's various side branches and ramifications. Of particular interest here is that the quarry face was once almost thirty feet in front of the position it occupied during Buckland's time. The earlier quarry face clearly intersected the cave entrance, as it has done ever since. It is as if the cave were a long hole poked through a loaf of bread, and that subsequent slices have been cut off, each one intersecting the hole. This means that the cave, and its contained fossils, must have been seen long before 1821. Back

The entrance to Kirkdale Cave, as illustrated in Buckland's first book, *Reliquiae Diluvianae.*

then, presumably, there were no curious observers like the medical gentleman, with sufficient knowledge to recognize the significance of the fossils. The age of enlightenment in such matters had not yet dawned.

Buckland's cross-section through the cave depicted a relatively thin stalagmite bottom, covered by a one-foot layer of mud. He correctly concluded that the mud had accumulated after the original stalagmite bottom had formed. The top of the mud was sealed by a second, thinner, layer of stalagmite. Thickening toward the sides of the cave, this continued up the walls. Stalactites hung from the roof. Most of the bones were found in the mud, and much of the overlying stalagmitic crust had been destroyed before Buckland's arrival, during their extrication. But some of the mud was still undisturbed. Fossils were quite plentiful in parts, and Buckland gave a colorful description of how the bones, coated with stalagmite, projected beyond the mud, like pigeon legs through a piecrust.

Buckland had never done any serious cave exploration before, though he had visited several caves on the Continent. But there he was, alone in the dark on all fours, with a cave full of bones that had to be excavated and interpreted.

Surprisingly, there were no complete skeletons in the entire cave. There were precious few complete bones either, most of them having been broken into small fragments. Some of the species still existed in Buckland's Britain:

fox, weasel, deer, rabbit, water rat, mouse, raven, pigeon, lark, snipe, and some ducks. But many belonged to exotic species, generally confined to warmer parts of the modern world: hyena, lion, bear, elephant, rhinoceros, hippopotamus, horse, and ox. What should he make of all this?

When he entered the cave, doubtless accompanied by some locals acting as guides, he expected to find evidence of the Noachian flood. The bones should have been swept and tumbled by the raging torrent, jumbling them together and mixing them with pebbles and stones, but this was not the case. However, the animals he found were clearly antediluvian because some of them, like the straight-tusked elephant and the narrow-nosed rhinoceros, were species Cuvier had shown to be extinct. Others, like the hippopotamus and hyena, no longer occurred in Britain. The enclosing mud must have been formed during the great flood. The most obvious interpretation was that all the animals had perished as a direct result of the inundation. The broken state of their bones would accord with the violence of the maelstrom. However, when he looked at the bones more closely, there was no evidence that they were water-worn: "[I] could not find a single rolled pebble, nor . . . one bone, or fragment of bone, that bears the slightest mark of having been rolled by the action of water."

Indeed, the broken ends of the bones still retained their sharp edges and delicately pointed processes. He concluded from this that the flooding of the cave by the rising waters must have been a gentle process. So, although the violence of the flood destroyed and remodeled the surface of the earth outside, inside the cave the contents were completely protected.

Buckland probably used a small trowel to dig into the mud, cleaning off each specimen with a brush and holding it up to the lamp for closer inspection. As he turned each new find over in his hands, he became aware of something quite extraordinary: Some of the bones were scored with a regular pattern of gouges. What possibly could have caused such marks?

At some point he realized the marks corresponded exactly with the spacing of the canine teeth of the hyenas in the cave: He had discovered a former hyenas' den. It was the hyenas, with their predilection for consuming bones, that had shattered almost every bone in the cave, including those of the other hyenas. As he learned afterward, this conformed to the habits of modern hyenas, which kill and consume sick and injured members of their own kind.

He noticed, too, that many of the bones were worn smooth on one side, whereas their complementary sides were completely unworn. Curved bones, like each half of the lower jaw, always had their convex sides worn smooth, but

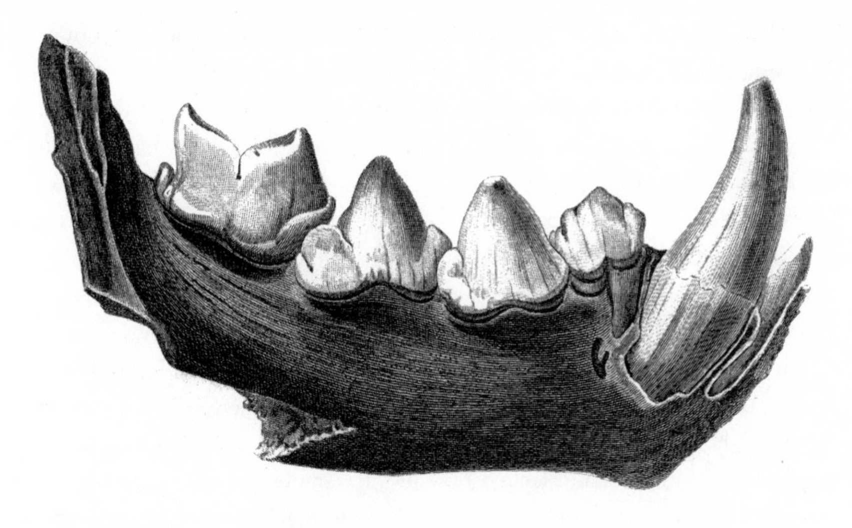

Part of the lower jaw of a hyena that Buckland collected from Kirkdale Cave.

their concave sides remained pristine. In an astute piece of deduction, he attributed these wear patterns to trampling by hyenas and to the rubbing of their fur on the side that lay uppermost on the cave floor. It was always the convex surface of a bone that was polished, Buckland reasoned, because convex-side-up is the only stable resting position of a curved object on a hard surface. Visualize an empty pistachio shell, or a peanut half, lying on a table. With the convex side down it is free to spin any which way, but hollow-side-down it is stable.

Buckland subsequently obtained some useful information from a big-game hunter who had once examined a tiger's den. The hunter had found a large stone on the floor of the den, on which the tiger habitually rested. This had been polished smooth by contact with the tiger's coat, lending indirect support to Buckland's explanation for how the cave bones became polished.

In addition to the bones of the hyenas' prey, Buckland found many small balls of calcareous excrement. When he showed these balls, called *album graecum,* to a menagerie keeper, the keeper immediately recognized them as hyena feces.

To confirm his interpretations Buckland, the consummate empiricist, kept a spotted hyena, named Billy, as a house guest for a short while. When given the shin bone of an ox, the hyena began to bite off large pieces, swallowing them whole as fast as he could. Buckland likened the power of its jaws to a miner's crushing mill. The hyena went on cracking the bone and extracting the marrow, licking out the condyle at the end of the bone, which he left untouched. The uneaten condyle was exactly like the ones found in the cave. The following morning Buckland examined Billy's *album graecum,* and found them to be identical to those from Kirkdale. Billy, incidentally, was only the forerunner of a veritable menagerie that would one day inhabit the Buckland household. There would be birds and snakes, sundry small mammals, a bear named Tiglath Pileser, an eagle and Jacko the monkey.

Buckland was impressed by the remarkable state of preservation of the cave bones. Unlike most fossil bones, they had not been mineralized, and retained much of the appearance of recent bones. He conducted a simple experiment to test their composition. Placing some of the fragments into acid, he dissolved out the mineral portion of the bone, and found that "nearly the whole of the original gelatine" had been preserved.

This simple experiment can be performed at home with some bones left over from a meal. When left in vinegar for a week or so the bones still look the same as before, but they feel as if they are made of rubber, and can even be tied in a knot. This is because the acid dissolves the calcium phosphate, the mineral portion that gives bone its hardness. What remains is the organic fraction, a protein called collagen, which gives bone its resilience. Boiling bones, as when making soup, converts the collagen into gelatin.

It is quite remarkable that the bones from Kirkdale had retained most of their collagen because the protein breaks down fairly rapidly, causing bones to become brittle. I have some bones of a domestic horse at my museum, dug up from a field in Ontario. They are probably no more than a hundred years old, but have already lost so much of their collagen that they are very brittle, and would shatter like china if dropped onto a hard floor.

The fresh condition of the Kirkdale bones, together with the limited growth of the stalactite in the cave, led Buckland to surmise that a relatively short time had elapsed since the mud had accumulated. Referring to estimates given by Cuvier and others for the universal deluge, he concluded that the cave deposits could not have exceeded 5,000 or 6,000 years. (Recent radiometric dating of the bones gives an age of 121,000 years.)

Buckland's investigations at Kirkdale were exemplary of the scientific method. They were in sharp contrast to the arguments he used just five years earlier, during his inaugural address as Reader of Geology. Had Buckland changed from subjective theologian to objective scientist during his professorship at Oxford? We could certainly draw this conclusion from much of his Kirkdale work. However, in the end, he let his biblical faith impede his scientific reasoning.

In an attempt to establish the "fact" of the Noachian flood, Buckland conjured up a rather imaginative scenario. He reasoned that the hyenas lived in the den right up until the time of the deluge, around 5,000 or 6,000 years ago. As the floodwaters rose, the hyenas rushed out, fleeing to the safety of the hills. The entire cave would have been flooded, preventing their return. He attributed the mud on the cave floor to the flood, although he noted that "Not a particle of mud was found attached either to the sides or roof." Should this not have struck him as odd? A common feature of floods is that the mud reaches as high up the walls as the water level. Would it not also have seemed strange for the hyenas' fecal balls to have remained intact after having been inundated by water for so long? And why should the sharp edges and delicate points of the broken bones have escaped the ravages of the floodwaters? It is difficult to imagine how the cave could have been flooded, then emptied again, without some fast currents being set up. This is especially so given the changes in elevation and the various side branches of the cave. The bones and feces would have been tumbled and abraded in the process, but there was absolutely no evidence for this. Furthermore, why should the bones be so well preserved, with most of their original collagen still intact, if they had been standing in water for the duration of the Noachian flood? Clearly, the evidence at Kirkdale failed to support a universal flood, but Buckland could not, or would not, see this.

Buckland was a man with preconceived ideas, thrust upon him both by the times in which he lived and by the theological environment in which he grew up and was educated. Although he was as capable of rational and objective thinking as any scientist today, he was selective in the evidence he chose to consider, and blind to that which he chose to ignore. Like many modern-day creationists, his agile mind was adept at performing the mental gymnastics necessary to reconcile religious beliefs with solid evidence. But, in contrast to contemporary times, Buckland did not have an extensive fossil record to weigh against the testimony of the Scriptures.

Buckland was probably apprehensive about how the results of his cave studies would be received, both by his theological colleagues at Oxford and by his geological circle in London. The first real test was his presentation of a paper to the Royal Society early in 1822. He was pleasantly surprised by the warm reception he received, and mentioned this in a letter to his friend, Lady Cole: "The president and Council of the R.S. have sanctioned my paper with the

Copley Gold Medal so that I am now not much afraid of any further opposition to my Hyena Story which my friends at first predicted no body would believe . . . "

He does not seem to have ruffled too many ecclesiastical feathers either, though he did receive some criticism for being inconsistent with the Scriptures. Critics held the view that the bones found in caves had been swept there during the flood from as far afield as the tropics. One cleric even suggested that the floodwaters could have shrunk the mouth of the cave after the bodies of the elephants and rhinoceroses had been swept inside.

Buckland's work at Kirkdale established his geological reputation, and he soon became one of the leading intellectual lights of the land. He received invitations to explore other caves, and one of these, on the Welsh coast, held an even bigger surprise for him.

Paviland Cave overlooks the sea on the exposed rugged coast of the Gower Peninsula, South Wales. There are two ways to reach the cave. One way, the sensible way, is to wait until low tide and clamber down the dog-toothed rocks of a deep gully to the beach below. It is then a short walk along the shore to a yawning gash in the rocks that marks the cave entrance. The unwise way, and the only way available at high tide, is to climb down a near-vertical rock face to reach the mouth of the cave directly.

Buckland learned of the cave from Lady Mary Cole and Miss Talbot during the winter of 1822. Both ladies, who were keenly interested in geology and frequently corresponded with Buckland, lived in Penrice, on the Gower Peninsula, close to Paviland Cave. He was keen to visit the new cave right away, especially since his book about caves, *Reliquiae Diluvianae,* was so close to completion, but he was prevented by other commitments. He eventually got there on January 23, 1823, and was accompanied to the cave by Miss Talbot and two others.

The small party no doubt took the safe route along the beach, and Buckland must have been pleasantly surprised by the large size of the cave. It is about fifteen feet wide and seventy feet deep, with such a high roof that it is possible to stand almost everywhere, except at the far end, where it narrows. The floor ascends fairly steeply, elevating the inner part above the rest, and it was here, in an earth-filled depression, that the bones were discovered.

Paviland Cave, on the rugged south coast of Wales, is only accessible from the seaward side. The top-hatted gentleman is taking the dangerous cliff approach.

Part of the skull of a woolly mammoth *(Mammuthus primigenius)* had been found before Buckland arrived, but it was poorly preserved and had been extensively damaged during its removal. Part of a tusk, about two feet long, was so decayed that the whole of its interior had crumbled to small fragments. But most of the fossils were still buried and awaiting discovery.

Buckland set to work, and soon began to find an assortment of antediluvian species, similar to those he had unearthed at Kirkdale. But he was in for a shock when he started digging near where the mammoth skull had been found: "I discovered beneath a shallow covering of six inches of earth nearly the entire left side of a human female skeleton."

Imagine his astonishment. Here were human remains, seemingly in a fossil state, lying alongside those of antediluvian creatures. But how could this be? It was a well established and sacred truth that humans, the pinnacle of God's creative powers, appeared on the Earth *after* the demise of these creatures. The great Cuvier himself had declared that human remains had never been found in a fossil state, and that all claims to the contrary had proved to be false.

So how did Buckland react to his startling discovery? Did his jaw drop as he took in the enormity of the situation? Maybe not. He tells us in his book, which went off to the printers within months of his visiting Paviland, that the entire cache of bones appeared "to have been disturbed by ancient diggings, and its antediluvian remains thereby have become mixed with recent bones and shells. . . . " The human remains, then, simply did not belong with the rest.

There were more surprises in store for him as he dug deeper.

"I found forty or fifty fragments of small ivory rods . . . and . . . some small fragments of rings made of the same ivory . . . nearly the size and shape of a small teacup handle . . . ."

There was also a piece of ivory about the same size and shape as a human tongue, some small seashells, and part of the scapula (shoulder blade) of a sheep. What did this all mean? To add to the mystery, the human skeleton, and all the ivory ornaments, were stained a dark brick-red. Buckland surmised from the latter that the body must have been daubed with the red pigment at the time of its burial. He chose to overlook the alternate explanation, that the staining was due to minerals, leaching in naturally from the surrounding soil.

The ivory ornaments had undergone the same amount of decomposition as the mammoth tusk. Buckland concluded from this that they must have been cut from the tusk, and at a time when it was still hard. His reasoning was perfectly sound, but, instead of drawing the logical conclusion that the human and mammoth remains were contemporaneous, he chose to believe that they were not. He correctly assumed the mammoth was of "very high antiquity . . ." but erroneously attributed a Roman age to the human remains, on the grounds that there was an archaeological site nearby. The Roman Britons, he reasoned, must have found the mammoth long after its demise. Here, as at Kirkdale, he chose to fit the evidence to his preconceived ideas.

Buckland was not only wrong about the age of the human skeleton—it has been radiometrically dated at 18,000 years—he was also wrong about its sex. He initially, and correctly, identified the skeleton as male. He had even toyed with the idea, perhaps not seriously, that he may have been an exciseman—no doubt romanticizing on the local stories of smugglers and wreckers, who were still active along this stretch of the Welsh coast. But he changed his mind, as he explained in a letter to Lady Mary Cole, written just weeks after visiting Paviland:

> The Man who we voted an Exciseman, turns out to be a Woman, whose History w[oul]d afford ample matter for a Romance to be entitled the Red Woman, or the Witch of Paviland—for some such Personage must she have been; but for what purpose she used her ivory Rods and Rings, & the shells in her Pocket I have yet to learn. . . . The Blade Bone of Mutton gives grounds for a conjecture, w[hic]h favors the Theory that she was a Dealer in Witchcraft. In Meyricks Hist. of Cardigan . . . are some curious Stories . . . of the Magical Powers of the Shoulder Blade of a Sheep . . .

And so was born the epithet, the Red Witch of Paviland, later changed to the Red Lady of Paviland to accommodate the sensitivities of the times. Buckland, the irrepressible wit, had a good deal of fun with the Red Lady too. He was a frequent recipient of poetry, both from students and colleagues, and shared the following with Lady Mary—left behind by one of his students:

> Have ye heard of the Woman so long under Ground
> Have ye heard of the Woman that Buckland has found
> With her Bones of empyrial Hues?
> O fair ones of Modern Days, hang down your head
> The Antediluvians rouged when Dead
> Only granted in lifetime to you

We cannot be sure whether Buckland was being entirely serious in claiming the human remains belonged to a witch. Not even his close associates knew when he was being serious, as Lyell recalled when describing Buckland's account of his exploits in the hyena den: Buckland, "in his usual style, enlarged on the marvel with such a strange mixture of the humorous and the serious that we could none of us discern how far he believed himself what he said."

There was no allusion to witches in his book, and the ivory objects and shells were briefly discussed as adornments, or possibly parts of a simple game. Nor was there any special mention of the possible significance of the sheep's scapula, which he included with the bones of other modern animals as the likely remains of human food.

Buckland's denial that the human species could be contemporaneous with extinct ones had ramifications for other investigations. The Reverend John MacEnery, who explored a cave in Devon two or three years after Buckland's Paviland explorations, found some flint tools beneath a thick unbroken layer of stalagmite. He reasonably concluded that the flints, hence their human users, must have predated the stalagmite, and were accordingly of great antiquity. However, as MacEnery wrote in his manuscript notes:

> Dr. Buckland is inclined to attribute these flints to a more modern date by supposing that the anc[ient]t Britons had scooped out ovens in the stalagmite and that through them the knives got admission. . . . Without stopping to dwell on the difficulty of ripping up a solid floor which . . . still defies all our efforts . . . I am bold to say that in no instance have I discovered evidence of breaches or ovens in floors . . .

MacEnery was unable to get his work published until after Buckland's death. An independent investigation of the Devon cave by another worker, Robert Austen, confirmed MacEnery's conclusions. Austen presented his findings in a paper read at the Geological Society in 1840. Significantly, the published summary of the paper made no reference to human artifacts. Similar fates met others who tried to publish papers that supported the coexistence of humans and extinct animals. But we should not hold Buckland solely responsible for the suppression of progressive ideas. Most people were simply not yet ready to reject the Genesis chronology, which placed the appearance of the human species after that of the other creatures.

Buckland's celebrity through his cave explorations kept him much in demand, as he mentioned in a letter to Lady Cole. He recounts dashing off from one end of the country to the other, in pursuit of his "prey," and teaching to overflowing classes.

He still managed to go fossiling at Stonesfield, however, and sometime between 1823 and the start of 1824, he made the discovery of a lifetime. This singular find would eclipse all his other achievements put together, but he, like Mantell, did not record the moment of discovery. His reaction is therefore a matter of speculation. Up to this point, recall, he had found some bones of a gigantic and unidentified creature, one that Cuvier surmised was a reptile that reached a length of forty feet. Buckland also knew that Mantell had been finding similarly gigantic remains in Sussex, but neither man knew what kind of reptiles they were dealing with. This was largely because the most informative part of the skeleton, the skull, had not been found. A skull can reveal a wealth of information, such as whether an animal was a carnivore or herbivore. It can also offer clues to an animal's sensory abilities—eyesight, hearing, and the like—and to the size and potential of its brain. A skull can also reveal an animal's relationships with others.

Unfortunately, Buckland did not discover a skull at Stonesfield. But he did unearth the next best thing: the major portion of a lower jaw, with several teeth still in place. This discovery, one of the greatest of the century, finally revealed the nature of the enigmatic giant bones he had found at Stonesfield. He would soon share that revelation with the rest of the world.

6

Revelation

Fossilists and social reformers alike could claim 1824 as a banner year—the dawning of a new era of understanding. At the same time Buckland was busy exploring caves and fossiling at Stonesfield, those preoccupied with the harsh realities of their present world witnessed Parliament's repeal of the Combination Act.

The Combination Act, which had been in force for the previous twenty-five years, made it illegal for workers to join together to improve working conditions, effectively banishing trade unions. The decision to repeal the act, thereby legalizing unions, was not owing to any largesse of the Tory government. It merely reflected the government's acceptance of the recommendations of a Select Committee of Parliament. Indeed, the prime minister, Lord Liverpool, was a staunch opponent of reform, and had introduced some sternly repressive measures during his twelve years in office. These included the Six Acts, rushed into the statutes in 1819, following the Peterloo Massacre.

This massacre, which took place in St. Peter's Fields, Manchester, was named in ironic allusion to the Battle of Waterloo. At Waterloo British troops had been used against the French, but at Manchester they were used against the English. A crowd of some 50,000 people had gathered in a peaceful demonstration for parliamentary reform, but the local authorities overreacted and sent in the troops. The soldiers charged into the crowd with swords drawn, killing 11 and wounding 400.

The Six Acts were aimed at preventing further demonstrations for social reform. They included such Draconian measures as prohibiting public meetings of more than fifty people, except with prior consent. Publications judged to be blasphemous or seditious were banned, and certain radical publications that had previously avoided paying stamp duty now had to pay. Stamp duty required every publication to pay a levy of four pence per copy, putting the cost of newspapers out of reach of all but the rich.

Peterloo and its aftermath spurred Shelley to pen "Men of England," one of several poems he wrote to rally the spirit of the working classes. Had he been living in England at the time, rather than in Italy, he would undoubtedly have been arrested for sedition—if not for the first verse, then certainly for the sixth:

> Men of England, wherefore plough
> For the lords who lay ye low?
> Wherefore weave with toil and care
> The rich robes your tyrants wear?
>
> Sow seed—but let no tyrant reap:
> Find wealth—let no impostor heap:
> Weave robes—let not the idle wear:
> Forge arms—in your defence to bear.

Mantell had been so appalled by the Peterloo Massacre that he had drafted a set of resolutions to the Prince Regent. He was no doubt delighted when the Tories struck down the Combination Act. But Buckland, lacking Mantell's zeal for social reform, was probably too busy with other things to pay it much heed. The most pressing engagement on his calendar for 1824 was his presentation of a paper on the Stonesfield giant to the Geological Society.

In preparation for the February meeting Buckland had engravings made of the newly discovered jaw, together with parts of the skeletons of several other individuals. Although he had nothing that even approached an entire skeleton, he hoped to give his audience sufficient information to enable them to recognize the creature, should they encounter its remains in their own collections. That way they might contribute to a more complete picture of the beast.

With a little over three weeks to go before his presentation, Buckland received some unexpected news from the West Country. The news was sufficiently important to cause him to drop everything and dash off to Lyme Regis, calling on his good friend Conybeare on the way.

The Reverend William Conybeare's passion for geology was second only to his devotion to God and to his church. He was Rector of Sully at the time, in South Wales, but lived in Bristol. His ecclesiastical duties kept him busy, but he still had some time for geology. Living in Bristol helped him keep abreast of the marine reptile discoveries being made in the West Country by Mary Anning, and by local collectors nearby.

Conybeare was very astute. Even before Anning's discovery of the world's first plesiosaur skeleton, which he subsequently studied and named in 1821, he had anticipated its existence. Not that he knew what kind of creature it was. All he knew was that a second kind of marine reptile coexisted with ichthyosaurs. This surmise was based on his discovery of some unusual vertebrae in a collection of ichthyosaur vertebrae belonging to a local fossilist. The unidentified vertebrae had relatively thicker centra (spools), and each centrum was permanently fused to its neural arch. He was eventually able to match the mystery vertebrae to those of Anning's first plesiosaur skeleton, confirming his suspicion.

The first plesiosaur skeleton, recall, had no skull, so Conybeare still did not know what sort of head plesiosaurs had. But sometime after its discovery, an isolated and incomplete lower jaw was found at Lyme Regis. The jaw was obviously not ichthyosaurian, and Conybeare suspected it probably belonged to a plesiosaur. Then a complete but badly crushed skull was discovered in the town of Street, partway between Bristol and Lyme Regis. Again, it was not ichthyosaurian, and Conybeare suspected that it too belonged to a plesiosaur, but he had no way of knowing. What was desperately needed was a complete skeleton, but Conybeare, in the meantime, had to make do with what was available.

Conybeare's studies of the available material convinced him that plesiosaurs formed a link between ichthyosaurs and crocodiles. Like Buckland, and most of the others in their geological circle, Conybeare thought that all organisms formed part of a "connected chain of organized beings" that God had created. Adherents to this belief were quick to deny any transmutational [evolutionary] connotation, and Conybeare made the point quite clear in his first paper on plesiosaurs: "When alluding to the regular gradation . . . the linked . . . series of animal forms, we would wish carefully to guard against the absurd and extravagant application which has sometimes been made of this notion [of transmutation]."

He went on to explain how every niche capable of supporting life had been filled, so there was a wide variety of different creatures, each providing:

> Striking proof of the infinite riches of creative design, [and] of the infinite wisdom . . . [of the Creator]. Some . . . however (and Lamarck is

The isolated lower jaw from Lyme Regis that Conybeare suspected was part of a plesiosaur (top), and the crushed skull from Street.

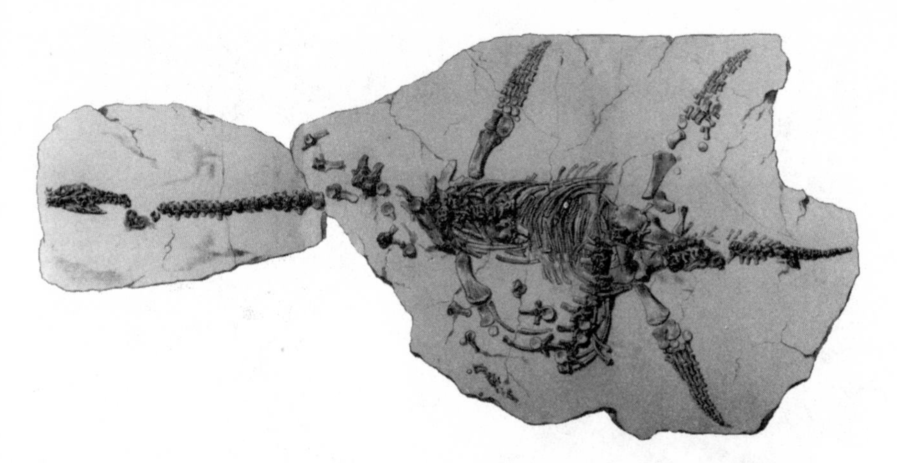

Mary Anning's second plesiosaur discovery—the first complete skeleton ever found. The specimen is on display at London's Natural History Museum.

> more especially censurable on this account) have most ridiculously imagined that the links hence arising represent real transitions . . . an idea so monstrous, and so completely at variance . . . with the evident permanency of all animal forms, that [it required] . . . nothing less than its bigotry to defend it.

Transmutation was clearly anathema to Conybeare.

It is unfortunate that Mary Anning had collected most of the first plesiosaur skeleton as separate bones, thereby losing their exact relationships. Conybeare's reconstruction of the fore paddle was accordingly conjectural, and the hind one was so incomplete that he did not even attempt a reconstruction. Some of the vertebrae appeared to have been lost during collection, so he could not be absolutely sure that the creature had quite such a long neck as he suspected. And, of course, the skull was missing. But this unsatisfactory situation was about to change.

Conybeare was busily working on a guest sermon he had to preach in Oxford when Buckland, en route to Lyme, came bursting in on him. Buckland, larger than life and doubtless in a great state of excitement, probably blurted his news: Mary Anning had just discovered a second plesiosaur! But this one, unlike the first, was a complete skeleton. Buckland told Conybeare that Anning had offered it to the Duke of Buckingham for £100. The duke had

Henry De la Beche, who collaborated with Conybeare on the study of marine reptiles, lived in Lyme Regis during his teenage years. He is often romantically connected with Anning, two years his junior, but there is no good evidence.

asked Buckland to examine the specimen and complete the transaction, if the skeleton was what it was purported to be.

Conybeare's peace and quiet was completely shattered by the visit, preventing him from getting on with his sermon, but he could not help being caught up in the excitement. He was anxious to know whether the new specimen confirmed his conjectures on the anatomy of plesiosaurs. "I begged him of course to send me immediate intelligence thence," Conybeare wrote his geological friend Henry De la Beche, who was currently in Jamaica, "and three days afterwards received a very fair drawing by Mary Anning of the most magnificent specimen. . . . You may imagine the fuss all this occasioned."

The Bristol Philosophical Society, which Conybeare had helped to establish, was meeting that very week, so he resolved to break the news there: "and to the Society I went, delighted . . . of making this strange monster first known to the public. . . . " Conybeare had only Anning's drawing to show his audience, but it was enough to demonstrate the singular features of the remarkable new discovery. "Such a communication could not fail to excite great interest," Conybeare told De la Beche. "Some of the folk ran off instantly (it was Friday evening) to Gutch's printing office [Bristol Press], whither I was obliged to follow to prevent some strange blunders . . ." being made by the journalists.

Buckland arranged for the specimen to be shipped to London, to be placed on temporary exhibition at the Geological Society. He charged Conybeare to be there to meet it, to make sure it did not fall into the hands of Sir Everard Home (1756–1832).

Sir Everard Home, first president of the Royal College of Surgeons, studied under John Hunter, the celebrated surgeon, who later married Home's sister. When Hunter died in 1793, Home became the executor of his will, and was entrusted with Hunter's unpublished manuscripts. Home was not noted for his productivity or brilliance, but he began to publish more papers than he had before. The suspicion naturally arose that he was plagiarizing Hunter's work. The suspicion was founded in fact, but the guileful Home destroyed the evidence by burning Hunter's manuscripts. Some of the geological circle knew of these nefarious deeds, which helps explain why he was not widely trusted or respected.

Buckland was to be elected president of the Geological Society at the forthcoming annual meeting, and hoped the plesiosaur would arrive in time for this auspicious occasion. So too did Conybeare, who was scheduled to deliver an address on its anatomy at the same meeting. Conybeare arrived in town as arranged, only to learn that the ship carrying the precious cargo had been delayed in the English Channel. Undaunted, he was determined to make the most of his unexpected trip to London. He stayed at the Salopian Coffee House, a hotel in Charing Cross Road that was Buckland's favorite haunt.

As it happens, the plesiosaur skeleton failed to arrive in time for the meeting, but Conybeare was not unduly concerned. He told his friend De la Beche that the annual dinner was:

> one of the pleasantest public meetings I have ever attended. Buckland as the new Pres.[iden]t was put to his oratory, & some dozen of us talked in our turn, but in place of the usual trash on such occasions every one had some interesting facts . . . to communicate. We adjourned to the Society's rooms at 1/2 past eight, & there I lectured on my Mstr [monster]. . . . .

Conybeare's presentation was an enormous success. He did not have the same flair for storytelling as Buckland, but he nevertheless painted a graphic picture of the bizarre anatomy of plesiosaurs. They were as alien to his audience as if they had hailed from another planet, invoking ethereal images of primordial seas ruled by wondrous reptiles. Using Anning's detailed drawing of the skeleton, he showed how the new discovery confirmed several of his conjectural points on plesiosaurian anatomy. The neck was as unbelievably long as he had predicted—longer than in any other animal. Initially he had counted twelve neck vertebrae, but the new specimen had about thirty-five. No other animal had so many neck vertebrae, the closest approach being birds, though he commented that they too fell short, with no more than twenty-three (some swans have upward of twenty-five). He went on to suggest:

> It swam upon or near the surface, arching back its long neck like the swan, and occasionally darting it down at . . . fish. . . . It may perhaps have lurked in shoal water along the coast, concealed among the seaweed, and raising its nostrils to a level with the surface from a considerable depth . . . a secure retreat from the assaults of dangerous enemies . . . the length and flexibility of its neck may have compensated for the want of strength in its jaws and its incapacity for swift motion through the water . . . .

Conybeare's image of a plesiosaur cruising along near the surface, with its long neck arched above the waves, became very popular, inspiring artists for

An early nineteenth-century vision of life in the sea during the Age of Reptiles, showing a swan-necked plesiosaur.

generations to come. But it was not very plausible. This is because plesiosaurian vertebrae, unlike those of birds, have restricted dorso-ventral (up and down) mobility, most of their flexibility being lateral (side to side). The notion of their lurking in the shallows with their nostrils raised above the surface is also implausible, not only because of the limited dorso-ventral neck flexibility, but also because of the problems of water pressure. When an animal is below the surface, water pressure tends to collapse its lungs, and this pressure increases with depth, by one atmosphere of pressure for every thirty-two feet (10 meters). The lungs therefore have to work against water pressure during inspiration, and this becomes increasingly difficult with increasing depth, which is why scuba divers must breathe compressed air.

Conybeare's studies enabled him to confirm that the isolated skull, found in Street a couple of years before, was indeed plesiosaurian, because the new skeleton had a similar type of skull:

> We now also learn for the first time, that the head of this animal was remarkably small, forming less than the thirteenth part of the total length of the skeleton; while in the Ichthyosaurus its proportion is one-

> fourth. This proportional smallness of the head, and therefore of the teeth, must have rendered it a very unequal combatant against the latter animal. . . .

The image of animals actively pitching into one another in mortal combat was a popular notion during Conybeare's time. De la Beche, for example, painted a watercolor depicting such a scene, sometime around 1831, and called it *Duria antiquior or Ancient Dorset*. The central figure, a large toothy ichthyosaur, bites through the long neck of a plesiosaur, and another plesiosaur launches a sneak attack from the shallows on a crocodile basking on the shore. The sea is a throng of biting, snapping jaws, and, in the distance, the long, disembodied neck of a plesiosaur launches itself into the air to snatch a low-flying pterosaur.

Such sanguine imagery of the ancient globe was based on a distorted view of the fossilists' nineteenth-century world. Today, thanks mainly to television, we know that animals in their natural habitat are not locked in eternal battles to the death. We know, for example, that combat between meat eaters, or between any other groups of animals, rarely becomes an overt conflict of tooth and claw. Rather, it is a subtle competition for resources and for living space. People in Regency England rarely had the opportunity to witness interactions between animals in the wild. Some individuals had seen lions and tigers in menageries, and could see how fierce they were. It was therefore natural for them to suppose that such animals spent most of their time in open conflict in the wild. It is not surprising, then, that the early fossilists were so obsessed with depicting combat among the denizens of the prehistoric world.

Conybeare told his audience that the limbs of plesiosaurs were as unusual as their necks. Their digits were exceptionally long, with upward of ten phalanges (finger bones), compared with only three in our own fingers. Such long, slender paddles were unique among the quadrupeds. Conybeare freely admitted that he got the humerus the wrong way around in the first skeleton, and confused the bones of the forearm (the radius and ulna). He also acknowledged that he had wrongly depicted the edges of the paddles as being formed of rounded bones, but this was because "when the specimen . . . was found, the bones in question were loose, and had been subsequently glued into their present situation, in consequence of a conjecture of the proprietor."

The "proprietor" referred to here, of course, was Mary Anning. He could have used the word "collector," but may have chosen the alternate word to convey the pecuniary, rather than collegial, nature of her relationship with the geological circle. The fact that he did not mention her name once during his entire address, even though she discovered both of the plesiosaur skeletons, substantiates his intent to exclude her.

Conybeare was criticized by some geologists at the time he proposed the name plesiosaur (meaning "approximate to the saurians") because the name did not refer to their unusually long necks. He responded to this in his presentation to the Geological Society by saying, "I think it very probable, from specimens which I have examined, that species of Plesiosaurus with shorter necks exist. . . . " He based this on differences in the relative proportions of their neck vertebrae. Conybeare was absolutely correct in this surmise because we now know that there are two distinctly different kinds of plesiosaurs. One group, today referred to as elasmosaurs, have long necks and small heads, like the first specimens he described. The other group, the pliosaurs, have short necks and large heads. That Conybeare could have made such a prophetic statement back in 1824, based on the very limited material available to him, underscores his outstanding abilities.

The evening of February 20, 1824, was a singular night in the annals of paleontology because the second speaker to take the floor was Buckland, who had an important discovery to announce. Buckland's presentation was riveting. Some of the audience had already heard of his earlier findings of gigantic bones at Stonesfield. But now they learned he had discovered more material of this remarkable reptile, including part of the lower jaw with teeth still in place. This was the vital missing part of his osseous jigsaw puzzle. Now that he had part of the skull, albeit only a small part, he was able to determine what sort of a creature it was.

Buckland, the irrepressible storyteller, beguiled his audience with his account of the terrifying beast. It had dagger-shaped teeth, six inches long, each rooted in its own socket. They were curved and pointed, with serrations along their cutting edges, like a steak knife. These were the teeth of a ferocious flesh-eater, and he dwelt upon the way the old ones were replaced by the new ones. The new teeth were formed in distinct pits beside the old, and he speculated how these may have been expelled by pressure exerted by the new ones. He emphasized the remarkable nature of this rapid tooth replacement.

Without giving reasons, Buckland told the assembled gentlemen that the vertebral column and legs resembled those of mammals, but that the teeth showed the animal to be reptilian. One of the mammal-like features that impressed him was probably the way the sacrum—the part of the backbone to which the hips are attached—was formed of five fused vertebrae. No other

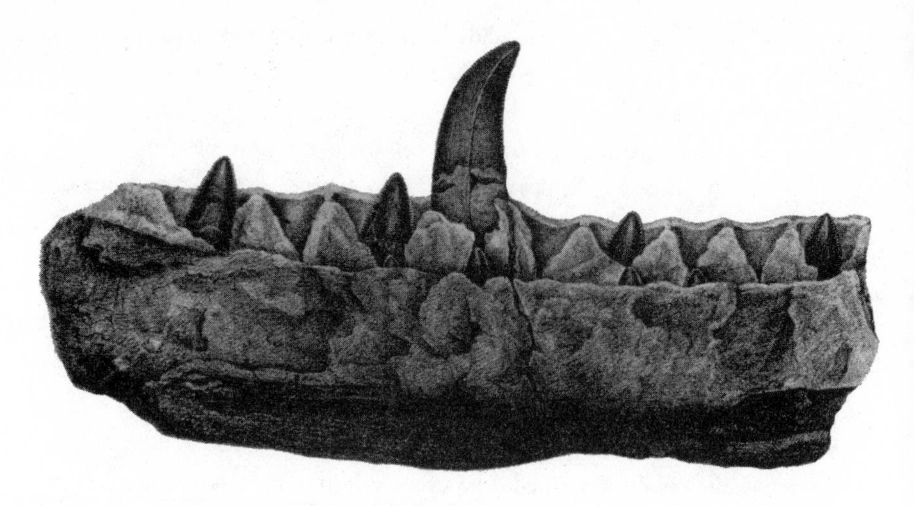

The lower jaw of Buckland's Stonesfield giant. The teeth, which are still in place, have serrated edges, like a steak knife.

reptile known at that time had such a sacrum. The limb feature that reminded him of mammals was probably the way the head of the femur—the ball that fits into the socket of the hip joint—was set off from the shaft, almost at right angles. In modern reptiles the head of the femur is in line with the shaft, causing the bone to be splayed at the sides of the body. The same is also true for the forelimbs. This gives reptiles their sprawling posture. Mammals and birds, in contrast, have an upright posture, where the femur (and humerus) is held vertically beneath the body. Despite its mammalian leg structure, the Stonesfield giant was clearly reptilian in the way its daggerlike teeth were continuously replaced, a feature not found in any mammal.

Buckland would have been in his element with such an attentive audience, and such a rattling good story to tell. His estimate of the size of the leviathan, based on the length of its limb bones, would certainly have captured their attention. The largest femur found at Stonesfield was two feet nine inches (84 centimeters) long, which was huge when compared to that of a modern lizard. Scaling up a lizard until it had an equally large femur gave a size estimate for his giant of over forty feet (twelve meters). A reptile of this size was hard to imagine, and Buckland may have paused here to let the information sink in. He then told his audience that Gideon Mantell had discovered an even larger femur in the Tilgate Forest deposits of Sussex. Buckland inferred, erroneously, that this gigantic femur belonged to the same species, and comparisons to a modern lizard gave a length estimate of sixty to seventy feet (eighteen to twenty-one meters). This information, delivered with the correct timing and cadence, might have evoked a collective gasp of incredulity from his attentive listeners. To appreciate his audience's reaction we should remember the sensation that large exotic animals, like giraffes and elephants, had caused at zoological gardens. Here was a considerably larger animal, it was also a reptile, and a carnivorous one at that.

The large size inspired the name he coined for the beast. This name, meaning giant lizard, was announced without fanfare: "I have ventured, in concurrence with my friend and fellow-labourer, the Rev. W. Conybeare, to assign to it the name *Megalosaurus*." Neither Buckland, nor any of the other gentlemen present realized it at the time, but he was announcing the name of the world's first dinosaur.

Megalosaurus was a most remarkable discovery—far more significant than the most exciting of today's paleontological finds. This is because it showed, very graphically, that giant reptiles roamed the land during the remote antediluvian past, just as they had inhabited the seas and the air. Some of the gentlemen may have been acquainted with the idea of the Age of Reptiles through their readings of Cuvier. But here was solid proof, in the skeletal remains he set out before them, of one of the leviathans from that distant age—an age when reptiles ruled the Earth, as mammals dominated the modern world.

None of his audience would have doubted that these monstrous reptiles were extinct, like the mammoths that came long after them. Nor would any theological toes have been trodden on by the revelation, because the Book of Genesis told of a former time when "There were giants in the Earth."

If any of the gentlemen were having difficulties coming to terms with the Age of Reptiles, it was probably because of their lack of imagination, and the absence of any points of reference in their own world. There were, for example, no modern representatives of marine reptiles because neither plesiosaurs nor ichthyosaurs left any living descendants. But Buckland painted a more tangible picture of *Megalosaurus*. He saw it as a scaled-up version of a modern lizard, referring to it as the "great fossil lizard of Stonesfield." He also thought it was capable of venturing into the water, probably because of its association with crocodile and turtle fossils.

Having introduced *Megalosaurus* to his audience, Buckland spent some time discussing the associated fossils found at Stonesfield, comparing them to Mantell's Tilgate fossils. The two faunas were remarkably similar. Both included terrestrial and aquatic animals, as well as terrestrial plants. And the two assemblages appeared to be formed under similar environmental conditions. However, Buckland pointed out that the thickness of the strata that intervened between the two formations ruled out the possibility that they were identical.

He was perfectly correct to assert that the Stonesfield and Tilgate assemblages were not the same. As noted earlier, they belong to the Middle Jurassic and Early Cretaceous, respectively. However, he was wrong to say that *Megalosaurus* occurred in the Tilgate fauna. Large carnivores similar to *Megalosaurus* were certainly known from Cuckfield, but the material was too incomplete to make a positive identification. Indeed, much of this material is still unidentified to the present day. He should therefore not have used the femur from Tilgate to estimate the size of *Megalosaurus*. His estimate of sixty to seventy feet was a gross exaggeration, by a factor of at least two, as we now know.

Although *Megalosaurus* was the largest and most impressive terrestrial fossil from the Mesozoic era, a tiny jaw had been discovered at Stonesfield that Buckland realized was "not less extraordinary." Buckland thought it was an opossum, because of its similarities with the living pouched mammals (marsupials). However, we now know that this primitive mammal, called *Phascolotherium,* was not a marsupial. What made it so special was that it was the first fossil mammal from the Mesozoic. Buckland, fully aware of the enormity of the find, told his audience that he identified the inch-long jaw as mammalian only on the authority of Cuvier, who had examined the specimen. He admitted that he would have hesitated to announce the unprecedented discovery of a mammal *below* the level of the chalk, had he not received the highest approbation of Cuvier.

The discovery was a major anomaly because it did not fit in with the accepted scheme of things. God had created different creatures at different times, beginning with the lowliest of animals, like fish, which first appeared in the older rocks. Life progressed through to the amphibians and reptiles, which first appeared at a higher level, to the mammals, which did not appear until later in the sedimentary series. The last mammal to appear in this continuous chain of beings was the human species, the last and greatest of God's creative works. So how could the presence of a mammal be explained in such ancient rocks?

Buckland did not offer any explanation at the time, and was quite content just to report the find, allowing Cuvier to take responsibility for authenticating the discovery. Others would sidestep the issue by arguing that the materi-

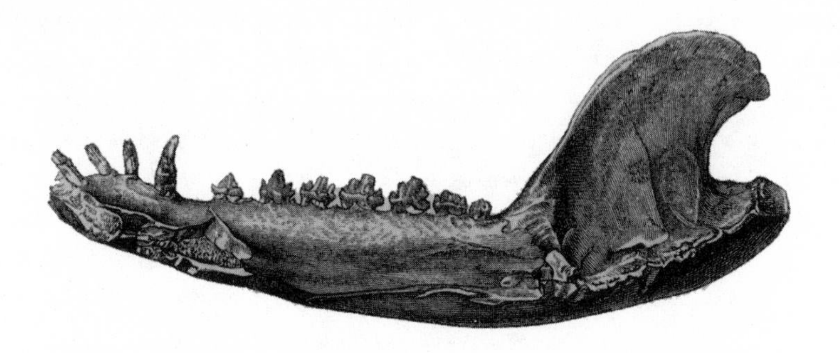

The Stonesfield "opossum" was an isolated lower jaw, only an inch (2.5 centimeters) long. This primitive mammal, called *Phascolotherium* ("pouched beast"), was the first mammal discovered from the *Age of Reptiles*.

al must have been reworked from the overlying, younger, Tertiary deposits. This was not a satisfactory solution, however, and the problem was discussed and argued for years to come. Buckland discussed the issue at some length in his second book, *Geology and Mineralogy,* published in 1836. This was one of a series of books in the *Bridgewater Treatises,* commissioned by the late Earl of Bridgewater, to show "the Power, Wisdom, and Goodness of God, as manifested in the Creation . . ." Buckland made the point that the molar teeth of the Stonesfield "opossum" had a double root, so there was no question it was from a mammal, but he reasoned it was a marsupial. Marsupials bore their young in an immature state, and shared a number of other "inferior conditions." They clearly occupied "an intermediate place between viviparous [live-bearing] and oviparous [egg-laying] animals . . ." forming "a link between Mammalia and Reptiles . . ." leading "us to expect . . . that the first forms of Mammalia would have been Marsupials." Buckland thereby removed the anomaly of discovering a mammal so early in the geological strata by interpreting it as a new link in the "grand continuous chain which connects all past and present forms of organic life, as parts of one great system of Creation." This, of course, was not an evolutionary chain, but one ordained by God.

Buckland's announcement of the Stonesfield mammal would have caused a great deal of interest on that dark February evening. So too would Conybeare's disclosure of the bizarre anatomy of plesiosaurs. But *Megalosaurus* had taken center stage. Some of the gentlemen in the room may have wondered whether *Megalosaurus* was the only giant reptile to stalk the land in those remote antediluvian times. There was, after all, more than one kind of sea dragon, so why not more than one kind of giant lizard? And although some may have speculated upon the matter, there were those present in the audience who knew, for sure, that other kinds of cold-blooded giants did exist: They had seen their fossilized remains with their own eyes.

7

The Tooth of the Iguana

The frontispiece of Mantell's second book, *Illustrations of the Geology of Sussex,* is an engaging pastoral scene of the quarry at Whitemans Green. Trees and shrubs encroach upon the quarry, and the summer sun casts short shadows on the warm rocks. Several quarrymen work away at a large boulder, under the watchful eye of two gentlemen in top hats, and a third top-hatted figure wields a mallet. A man in the background pushes a wheelbarrow across a narrow plank, spanning a gully at the top of the quarry. And away in the distance, beyond the man and beyond his wheelbarrow, stands the spire of Cuckfield Church. The quarry was abandoned and filled in long before Mantell collected his last fossil. But if you stand on the cricket field that is now Whitemans Green, and peer through the trees at the Cuckfield Church spire in the distance, you can see where the quarry used to be.

During the early spring of 1824 Mantell, Buckland, and Lyell were rained out of the quarry. This was a great pity because Mantell's two visitors had come from London expressly to visit Cuckfield and to see his Tilgate fossils. Mantell had seen Buckland just three weeks before, when *Megalosaurus* had been officially named. Soon after that meeting Mantell had sent him a generous sample of Tilgate fossils. Buckland wrote straight back, telling him "that you have also the *Megalosaurus* in your neighborhood. . . . " Once Buckland had a chance to examine Mantell's extensive collection of Tilgate fossils, he could see for himself that the physician had a veritable charnel house of giant lizards. Unfortunately, the material was rather incomplete, like his own Stonesfield collection, with far more bone fragments and broken teeth than whole specimens.

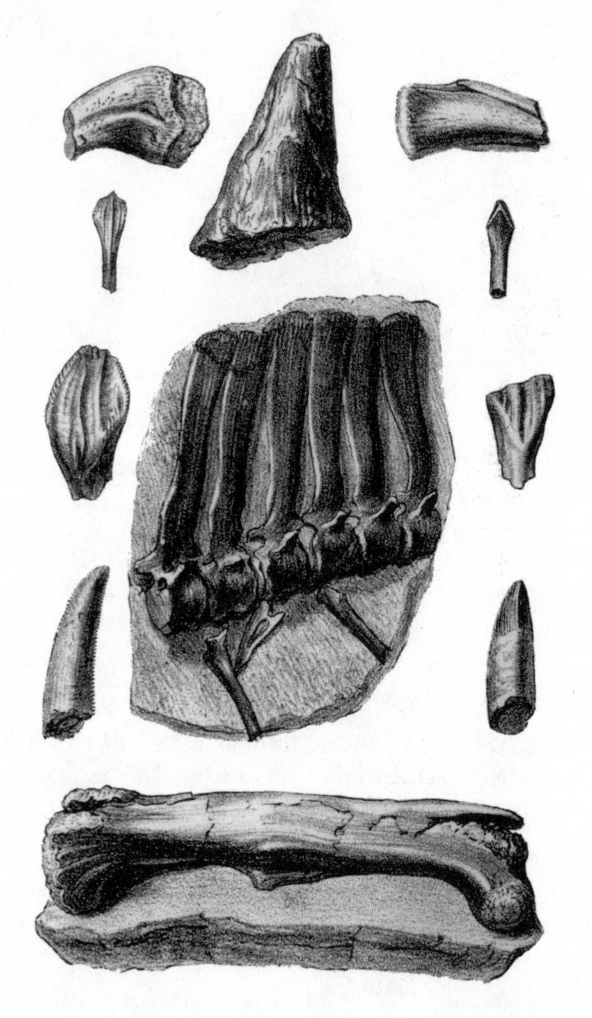

Some of the Tilgate fossils from Mantell's collection (not drawn to the same scale). Except for the two pointed teeth above the long bone (femur), they all belonged to Mantell's giant reptile with the fluted teeth.

Some of the bones were massive, far exceeding the size of Buckland's *Megalosaurus* specimens. Buckland thought they were whale bones, but we know today that whales did not appear until after the Mesozoic Era, long after the demise of *Megalosaurus* and its reptilian kin. These bones may have belonged to sauropod dinosaurs. Buckland was inclined to assign most of the large bone fragments to *Megalosaurus,* but Mantell was not convinced. Neither was Mantell's friend Lyell. Soon after his first visit to Mantell's collection, Lyell had raised the question whether the Stonesfield and Tilgate giants might belong to different kinds of reptiles. It was not that Mantell questioned the occurrence of *Megalosaurus* in his Tilgate collection—he had found some daggerlike teeth with serrated edges that he thought could hardly belong to anything else. But he felt that most of the large bones belonged to a different kind of giant reptile, the one that had left behind so many of the fluted teeth that Cuvier had identified as rhinoceros.

The three men probably discussed the fluted teeth at great length. Mantell lacked Buckland's extensive experience of fossils, and he could not hold a candle to Cuvier's vast anatomical knowledge. Nonetheless, he still thought the fluted teeth belonged to some new kind of giant reptile. Lyell's interests lay more in rocks than fossils, but he was inclined to agree with Mantell. Buckland, however, remained convinced that the mystery teeth were not reptilian at all, and probably belonged to some kind of fish. The only alternative, to Buckland's mind, was that the teeth really did belong to a rhinoceros, and that they had been introduced into the Tilgate strata from the overlying diluvial deposits.

Buckland wrote several letters to Cuvier after his trip to Sussex. He drew a sketch of the fluted teeth in one of them, so there could be no misunderstanding to which teeth he was alluding. "Mr. Mantell has certainly the bones and teeth of *Megalosaurus* in Tilgate Forest," he wrote, "but the teeth of which you speak [the fluted ones] are not of *Megalosaurus* but I think . . . part of the jaw of some fish. . . . " Buckland wrote to Mantell the same day, telling him, "Cuvier is much puzzled about your large [fluted] . . . teeth. I think they must have come from some fish like *Diodon* or *Tetradon* [puffer fish]. . . . "

Enigmatic teeth were not the only thing on Buckland's mind during the summer of 1824. During a trip to the West Country he was coaching through Dorset, reading Cuvier's most recent volume, when he noticed that one of the other passengers, a young lady, had the same large tome among her books. When Buckland commented on the fact she told him she had drawn the illustrations for Cuvier's book. At this, Buckland is reported to have said, "You must be Miss Morland, to whom I am about to deliver a letter of introduction."

Mary Morland, whose mother died when she was an infant, had been raised by the noted Oxford physician, Sir Christopher Pegge, and his wife. Buckland knew Pegge, having attended his lectures during his student days. Presumably Morland was taking a holiday in Dorset, and Pegge, knowing this was also

Buckland's destination, had given him a letter of introduction to her. Their common interest in fossils soon blossomed into romance, and the forty-year-old bachelor married her the following year. Things always seemed to fall into Buckland's lap, because, shortly before his marriage, he was appointed Canon of Christ Church, Oxford. The title carried no duties, but provided him with a fine house and an annual stipend of £1,000—a considerable sum in those days.

Mantell, in contrast, had to work hard for everything he achieved. Whereas Buckland had received an instant response from Cuvier on his Stonesfield giant, Mantell had to bide his time to receive the great anatomist's final word on his Tilgate monster. But, at length, Cuvier replied. The letter, in French, was written three weeks after Cuvier received Buckland's letter, with his drawing of the fluted teeth:

> These teeth are certainly unknown to me; they are not from a carnivorous animal, however I think that they belong . . . to the order of Reptiles. From their external appearance you could mistake them for fish, similar to tetrodons or to diodons; but their internal structure is greatly different from these.

Cuvier clearly thought Buckland was mistaken in his identification. He then went on to ask the rhetorical question:

> Is it possible that we have here a new animal, a herbivorous reptile? . . . Time will confirm or invalidate this idea, since . . . one day we will . . . find part of the skeleton reunited to portions of jaws bearing teeth. It is that last object, mainly, that we need to look for with most perseverance.

Mantell must have been overjoyed when he read this letter because it confirmed his suspicion that he had discovered a new kind of reptile. But it was a reptile completely unlike any other kind—a reptile that masticated plants, like a cow. He wrote back to Cuvier (in English) expressing his deep gratitude, and sending some more teeth for his examination. He also included some illustrations from his forthcoming book, *Illustrations of the Geology of Sussex.* Mantell ended his letter with an apology for "the hasty and almost illegible manner in which I have addressed you; but I write under the pressure of a severe indisposition which has confined me to my room some time. . . . "

Mantell, who was only thirty-four at the time, did not enjoy good health. His problem, which dogged him for the rest of his life, stemmed largely from severe pains in his lower back. He would later attribute this to bending over his youngest daughter's bed during the protracted illness that began during her adolescence, and to an injury he suffered in falling from his coach. Many years later, however, a postmortem examination would reveal the cause to be scoliosis. The extreme lateral curvature of his lumbar region was accompanied by a marked rotation of several of the lumbar vertebrae. As a consequence, the transverse processes on the left side of these vertebrae, which normally project laterally, were directed backward—they could be felt as hard lumps in the small of his back. Medicine was in such a primitive state in those days that the orthopedic specialist he consulted completely overlooked the curvature of his spine. Moreover, he misdiagnosed the displaced transverse processes as tumors. But Mantell must have compensated for the deformity in his posture because a contemporary described him as being "tall, graciously graceful, and flexible."

Mantell, greatly inspired and energized by Cuvier's endorsement, bent to the task of determining what sort of new reptile he had discovered. Imagine the enormity of his task. All he had to work on were some fragmentary limb bones, some vertebrae, and some isolated teeth, probably all from different individuals. He had noticed from the outset that the teeth had wear surfaces rather than sharp edges—sheeps' teeth rather than lions'—and realized that they belonged to a herbivore. But what sort of herbivorous reptile was it? A singular feature of the teeth was that their margins were finely fluted. He had never seen teeth like these in any of the modern animals he had examined. But he needed to see much more material. To that end he visited the Hunterian Museum of the Royal College of Surgeons, in London, which housed one of the largest collections of extant skeletons in the country. Mr. Clift, the curator, a "bright little bald-headed man" ransacked all the drawers containing reptilian jaws and teeth, but without finding a match. Then a young assistant, Samuel Stutchbury, drew Mantell's attention to an iguana lizard he had recently finished skeletonizing. Mantell must have experienced one of those rare endorphin rushes when he saw the skeleton, because there, right before his eyes, were teeth strikingly similar to the fluted ones from Tilgate. They were considerably smaller, but the resemblance in shape was unmistakable. Mantell was elated. Here, at last, was a living analogue of the fossil that had been puzzling him for the last several years.

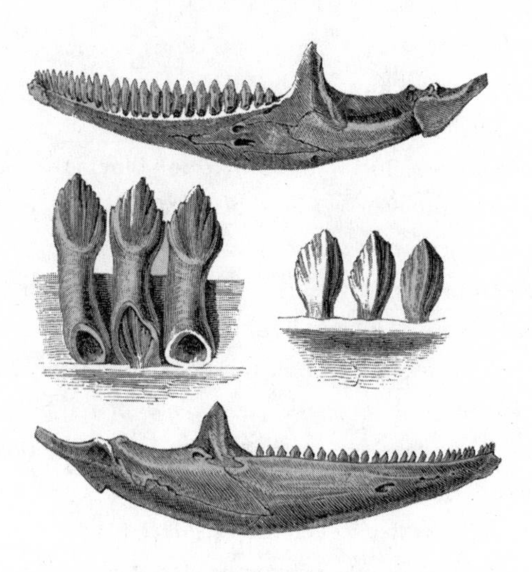

The lower jaw of the modern iguana lizard, showing details of the teeth, as illustrated by Mantell. The teeth are strikingly similar to the much larger ones Mantell found among the Tilgate fossils.

Mantell continued collecting more of the fluted teeth, making detailed comparisons with the modern iguana. He also began writing up a description of his new reptile for publication. He had to find a suitable name for the new fossil, a matter he did not take lightly. He wrote both to Cuvier and Buckland to seek their opinions. "I have ventured to propose the name *Iguanosaurus* for the fossil animal," he told Cuvier, "as indicating the resemblance it bears to the recent ones." In addition to the fluted teeth, Mantell obtained some more bones. These included an unusual spike-shaped bone, several inches long, reminiscent of a horn. He probably mentioned this in his letter to Buckland, because Buckland referred to the bone in his reply, saying that ". . . I . . . see myself no objection to your name of *Iguanosaurus*. . . . I am very curious to see your rhinoceros-like horn and wish you could either send it here to be exhibited or come up with it yourself to our next meeting. . . ."

Buckland went on to ask Mantell if he had any duplicate *Megalosaurus* bones that he could compare with his Stonesfield material. He was especially interested in the largest specimens because one of Mantell's Tilgate femora was nearly double the size of anything he had found.

Mantell's paper on *Iguanosaurus* was read before the Royal Society on February 10, 1825. Shortly after, he received a rather officious letter from Conybeare, whom he had not yet met, criticizing him for the name he had chosen: "Your discovery of the analogy between the Iguana & the fossil teeth is very interesting—but the name you propose *Iguanosaurus* will hardly do. . . . "

Conybeare's objection was that the name, meaning "iguana lizard" was equally applicable to the modern iguana. Among his alternate suggestions was *Iguanodon,* meaning "iguana tooth." Conybeare was held in such high regard that Mantell gladly accepted his advice, changing the name to *Iguanodon.* At about the same time, Mantell received a letter from Buckland concerning the horn-like bone. Referring to the horny protuberances some modern lizards bore on their snouts, Buckland suggested it might, indeed, have been a horn.

This seemed a very reasonable suggestion to Mantell, and, many years later, when enough *Iguanodon* material had been collected to reconstruct a skeleton, the animal was given a single horn on the tip of its snout. It was only during the latter part of the nineteenth century, when some remarkably well preserved skeletons of *Iguanodon* were discovered in a Belgian coal mine, that it was realized that the "horn" was really the fused terminal bones of the thumb.

Mantell's paper announcing the discovery of *Iguanodon* was published in the prestigious *Transactions of the Royal Society*. It was just as true in 1825 as it is today that although some publications slip into oblivion, others are an instant success, heralding the author in a blaze of light. Such was the case with the *Iguanodon* paper. Mantell was celebrated and acclaimed, rocketed into the cosmos alongside the other luminaries of the geological circle. "I shall ride on the back of my *Iguanodon* into the temple of immortality!" Mantell wrote in his journal, repeating a prophecy expressed to him by Robert Bakewell, the author of one of the most influential geological books of the day.

Mantell was elected a Fellow of the Royal Society toward the end of the year. "It was with no small degree of pleasure," he recorded in his journal, "that I placed my name in the Charter book, which contained that of Sir Isaac Newton and so many eminent characters."

Letters of congratulation flooded into Castle Place, his home. So too did a succession of visitors, some very eminent, and all to see his now famous collection. Mantell's journal entry for the end of the year (1825), which was usually so pessimistic, was remarkably positive, aside from his health concerns:

> The close of a year that has been so fruitful . . . I have advanced my literary reputation, and have been elected an honorary member of the Institute of Paris, and a Fellow of the Royal Society of London. My practice has considerably increased, but my strength and health have more than decreased in proportion.

Mantell's fossiling and geologizing continued apace during the next few years, alongside all his other responsibilities: the busy medical practice, his growing museum collection, his wife and family, and his friends and visitors. He sometimes managed to combine family life with fossiling by taking one or other of his family members along with him on his geological trips. Seven-year-old Walter was taken on a trip to Cuckfield quarry, shortly after his mother gave birth to her fourth and last child, Reginald, in 1827. Walter's eldest sister, Ellen, had an even greater adventure the following spring. An enormous bone was reported weathering out of the chalk cliffs, near Brighton, and her father took her with him to excavate it.

"After three hours hard labour I succeeded in laying bare a bone 30 inches in circumference, and 9 feet long; a fragment of either the rib or jaw of a whale." It was a remarkable discovery, but disaster was close at hand. "In attempting to remove it it fell into a hundred pieces! A few fragments were the only relics I could bring away of this, the most magnificent fossil I ever discovered."

His lost prize was obviously not part of a whale, but may have been the remains of a large plesiosaur, or perhaps a mosasaur. Mosasaurs were marine reptiles, closely related to living monitor lizards, and were the first giant reptiles to receive scientific attention. The first specimen caused a sensation when it was discovered in 1770 in an underground chalk cavern near the Dutch town of Maastricht. All that remained was a massive jaw, but Cuvier's studies revealed it belonged to a giant extinct monitor lizard. Although this accords with our modern view of mosasaur affinities, the relationship between the different types of reptiles was not altogether clear back in those days. For example, Buckland tells us in his *Bridgewater Treatise* that mosasaurs were an intermediate link between monitor lizards and iguanas, whereas *Iguanodon* and *Megalosaurus* were lizards. However, as we know today, monitors and iguanas are merely different types of lizards.

Mantell's loss of the giant bone is attributable to the unsophisticated collecting methods of those times, exacerbated by the natural brittleness of bones found in chalk. The specimen probably remained intact while he removed the overlying chalk because it still had the mechanical support of the underlying rock, but when he tried to remove the rest of the chalk the specimen disintegrated. Today the whole bone would have been thoroughly strengthened with plastics, then jacketed in burlap and plaster, before attempting to remove it.

Mantell also suffered some fossiling setbacks at the hands of others. Visiting Cuckfield quarry in December of 1830, he endeavored to obtain some fossils from the quarrymen with whom he had dealt for many years: "[But] the ungrateful scoundrels refused to let me have one, having found a customer on the spot; a gentleman who has just become amateur and collector; here there is an end to all my hopes of discovering the jaw of the *Iguanodon*."

The culprit, one Mr. Trotter, had presumably offered the quarrymen more than they expected to receive from Mantell. Trotter then added insult to injury by calling on Mantell the following year and presenting him with a cast of a magnificent tibia, which, Mantell recorded, "ought to have been mine." Later on that year Mantell visited Cuckfield and collected a few fossils, but "all the best [were] poached by Mr. Trotter." In spite of the hostility he harbored for him, they continued calling upon one another: "Walter and I drove to Cuckfield and visited Mr. Trotter: saw some magnificent fossils he had poached from the Cuckfield quarries!"

Mantell had struggled all his life to overcome the disadvantages of his modest beginnings, and in 1830 his craving for recognition and for the attention and favor of the aristocracy was about to be satisfied. Following the June 26 death of the much despised King George IV, King William IV ascended the throne. In marked contrast to his profligate brother, the sixty-four-year-old, who nobody ever dreamed would one day be king, was a sensible, pragmatic man. One of the first visits the popular monarch made was to Lewes, as Mantell recounts in his journal: "Their Majesties had signified their intention of seeing my Museum . . . but the various ceremonies occupied so much time that it was too late . . . a message was sent . . . that their Majesties would honor me another opportunity. I was presented to His Majesty. . . . "

Early the following year Mantell received a visit from the royal princes. The two boys took a great deal of interest in his fossil collection, and stayed for almost three hours. But all this celebrity was not without cost, primarily in the coinage of his precious time. "Worried to death with visitors," he wrote, following an overnight visit from Lord Viscount Cole, "really this notoriety is a great curse after all."

Mantell still continued to court the nobility, and spent increasing amounts of time in the fashionable environs of Brighton, pursuing "several of the first families in this part of the County." He contemplated relocating to Brighton, but wrestled with the wisdom of such a move: "Shall I leave this dull place and venture into the vortex of fashion and dissipation at Brighton, or shall I not? Prudence (with four children) says stay where you are—but ambition and my friend Martin Crispps Esq. says go and prosper! What shall I do?" He decided to stay in Lewes for the time being, but he would eventually move to Brighton at the end of 1833.

For all his attempts to ally himself with the nobility, Mantell remained a fierce defender of the common man, railing against injustice: "Went to Church with Ellen and Hannah and heard the Assize Sermon—a wretched discourse on passive obedience and non-resistance and all that. What a period for such humbug." The assizes that the sermon referred to were the sessional courts that periodically traveled the country, administering justice—or the nineteenth-century version of justice:

> The Assizes terminated yesterday; two poor fellows are to be hung, and many to be transported, or imprisoned. It is all bad; our peasantry are in a state of positive ignorance and slavery; almost starving, without the knowledge necessary to enable them to attempt obtaining redress without violating laws, which are made to oppress the poor and protect the rich!

Mantell did not exaggerate the gross injustices in Britain during these times of food riots and demands for political reform. Women, of course, had no vote, and only wealthier men had the right; universal suffrage did not come to Britain until 1928. The unrest even reached out and touched Mantell in the tranquillity of his rural Sussex. One dark November night, just after midnight, he was awakened from a deep opiate sleep, induced to control his pain, by loud cries of fire and the ringing of the fire bell. He jumped out of bed and saw the ruddy glow of burning barns and haystacks. He quickly dressed and hurried to assist.

The fires had been deliberately set, and Mantell organized some men to pull down a fence to prevent it from spreading. He then led a search for other fires. He returned home, exhausted, at five in the morning, having set the broken leg of a firefighter who fell from a haystack. Similar fires occurred in other parts of Sussex and in adjoining counties. Special constables were sworn in and night patrols were established. There was talk that the arson was the work of a political party, but Mantell believed it was "more likely to be effected by the peasantry, who have for years been ground to the earth by their masters. . . . Landlords will not see the necessity of calling the farmers together and making arrangements for their paying the peasantry in an adequate manner!"

The most singular instance of Mantell's crusade for the disenfranchised concerned the remarkable case of Mrs. Hannah Russell. Accused of poisoning her husband, she was put on trial with her alleged accomplice, their nineteen-year-old lodger, Daniel Leney. Mantell, who often attended the Assizes, was at the trial. He found so much fault with the prosecution's expert medical witness that he scribbled a note to her lawyer, posing a series of questions that should be asked about the postmortem. Unfortunately, he was called away to see a patient, and when he returned, some hours later, the trial was already over. Both of the accused had been found guilty and sentenced to death.

Hannah Russell's execution was delayed while certain legal issues were resolved regarding her marital status. Meanwhile, Mantell contacted the High Sheriff, outlining the medical and chemical errors in the prosecution's case. Working with Hannah Russell's lawyer, Mantell compiled a weight of evidence to show that her husband died from a heart attack, not from arsenic poisoning. Hannah Russell was eventually released and pardoned, as a direct result of his intervention. But this all came too late for Daniel Leney. He was hanged just a week after his trial, along with a laborer who stole two cows, a chimney-sweep who stole some sovereigns, and a sawyer who stole some silver. Bad science was as lethal in the courtroom as it was in the physician's surgery.

8

Just Causes

While Mantell was using his spare moments in the service of the law, his friend Charles Lyell—a lawyer—was devoting increasing amounts of his time and advocatory skills to geology. Lyell and Mantell shared similar geological views, both holding that geological phenomena, from the raising of mountains to the carving of valleys, could be explained by ordinary causes, without invoking the hand of God. They were good friends, and kept in regular contact through their letters. Whenever Lyell got the chance to get away from London, he and Mantell would spend time together, looking at rock sequences in Sussex.

Lyell sought rational explanations in causes that were still operating in the modern world. Lyell's steady-state view of Earth history, devoid of catastrophes like the Noachian flood to explain geology, is often referred to as *actualism* or *uniformitarianism*. The London barrister did not claim originality for the concept, which was first proposed in 1788 by James Hutton (1726–1797), a Scotsman who initially studied medicine, but who never practised. But Lyell was the first to amass such a weight of supporting evidence for uniformitarianism. He did this by putting his legal career on hold and traveling extensively throughout Europe and the British Isles, gathering hard evidence from the rocks.

Lyell's geological views were the very antithesis of those of Buckland, who, we have seen, sought to explain much of the past in terms of the Noachian flood. Lyell's opposition to such diluvial interpretations often pitted him against his old teacher, though it was often Buckland's confidant, Conybeare, who led the charge. The venue of their exchanges was usually the Geological Society. Mantell's professional commitments usually prevented him from

Charles Lyell, who became the most influential geologist of his time.

attending the bimonthly meetings, so Lyell kept him up to date through his letters. "A splendid meeting last night," Lyell wrote in the spring of 1829.

> Conybeare's paper on Valley of Thames, directed against Messrs. Lyell and Murchison's former paper, was read in part. Buckland [was] present to defend the "Diluvialists" as Conybeare styles his sect, and us he terms "Fluvialists" [from river]. . . . Murchison and I fought stoutly, and Buckland was very piano [subdued]. Conybeare's memoir is not strong

> by any means. He admits three deluges before the Noachian! and Buckland adds God knows how many *catastrophes* besides . . .

Lyell recounted another lively exchange, several weeks later:

> The last discharge of Conybeare's artillery, served by the great Oxford engineer [Buckland] against the Fluvialists . . . drew upon them . . . a sharp volley of musketry from all sides, and such a broadside at the finale from Sedgwick [Buckland's equivalent at Cambridge], as was enough to sink the "Reliquiae Diluvianae" for ever, and make the second volume shy of venturing out to sea.
>
> [Buckland never did publish his expected second volume on caves.]

Lyell never missed an opportunity to attack Buckland's diluvial theory. This was not because he was against religion, which he was not, nor against the man. But he was vehemently opposed to Buckland's use of his academic position and lofty scientific reputation in his attempt to show that geology conformed to the Scriptures. As far as Lyell was concerned, Buckland was misusing his considerable influence to the disservice of geology.

Metaphorical sparks were flying at the Geological Society, but real sparks—and smoke and steam—were being raised at Rainhill, near Liverpool, in England's industrial north. The directors of the recently formed Liverpool and Manchester Railway Company, unsure whether to use stationary engines or steam locomotives to haul carriages between the two cities, conceived of a locomotive competition to help them decide. A prize of £500 would be awarded to the winning locomotive that met all their specifications. These included maintaining an average speed of ten miles per hour and drawing carriages that totaled three times the weight of the locomotive, which itself could not exceed six tons. The locomotives had to run twenty times up and down the nearly two-mile stretch of test track, approximating the distance the trains would have to travel between the two towns when the railway was built.

Some 10,000 spectators showed up for the first day of the trials, waging bets and enjoying all the festivities of the racetrack. Ten entries had been received, but only five locomotives turned up.

This delightful scene, sometimes mistaken for the *Rocket* at the Rainhill, shows a very similar locomotive—probably the *Northumbrian*—that was built shortly afterwards.

Cyclops was disqualified when the judges discovered it was powered by a horse walking on a drive belt. *Perseverance* was damaged in transit and its inventor had to spend several days trying to repair it. He was forced to withdraw when his battered locomotive would not exceed six miles per hour. *Sans Pareil* had a good start, and reached a top speed of sixteen miles per hour, but a cylinder cracked on its eighth run, forfeiting the competition. The flimsiest entry, the *Novelty,* weighed only two tons and won the hearts of the crowd when it reached twenty-eight miles per hour on the first day. It suffered an overheated boiler pipe on the second day, however, necessitating a makeshift repair, and when the locomotive reached fifteen miles per hour the pipe failed catastrophically, causing extensive damage.

The final entry, Stephenson's *Rocket,* completed the trial without mishap, at an average speed of twenty-four miles per hour, easily winning the prize. The secret of the *Rocket*'s success was in its unique boiler design, where the heat from the furnace was transferred to the water through a battery of pipes. The drastic improvement in efficiency ensured an ample supply of steam to the cylinders, enabling high speeds to be maintained. Steam locomotives had been in existence for more than two decades, but George Stephenson's (1781–1846) revolutionary design heralded the new era of steam locomotion.

Lyell may have watched all this with a certain amount of detachment: rail and steam might be about to transform the world, but he was at work on more esoteric problems. He was interested not in how industrial forces were changing the landscape, but in how natural forces that were reshaping the land could be used to interpret the past.

Lyell had been working on a book, *Principles of Geology,* during the years leading up to the Rainhill Trials. His goal was to persuade people that geological changes that had taken place over aeons of time could be explained in terms of processes still taking place in the world. The first volume of the book was published in 1830, and in the spring of that year, as the manuscript neared completion, Mantell anticipated its likely impact: "the saints will be in an uproar. Yet he [Lyell] has taken great pains to avoid hurting the feelings of any while pointing out facts which so tremendously erase some orthodox opinions."

Lyell had originally planned to write an introductory-level book, hoping it would be sufficiently popular to generate significant earnings. Ideally the royalties from the book, taken with the annuity from his father, would allow him to give up the law and devote all his energies to geology. But, as he began documenting his case with the mass of evidence he had accumulated, the work expanded into a scholarly treatise of three volumes. The advocatory skills of his profession, matched by his ability to write popular prose, enabled him to extend beyond the geological circle and reach that intellectual segment of the populace who so craved knowledge of the new sciences of geology and paleontology. The book was an immediate success.

The second volume was published in 1832. This was to have completed the work, but it took much longer to write than anticipated, so it was decided to publish the last section as a third volume, the following year. By that time Lyell had already revised the first two volumes, their second editions appearing in print in 1832 and 1833. *Principles* enjoyed a remarkably long life, and Lyell was kept busy revising new editions of all three volumes for the rest of his life.

Darwin took a copy of the first volume of *Principles* on his five-year voyage around the world aboard the *Beagle.* In later years he wrote that his first geological foray of the trip "convinced me of the infinite superiority of Lyell's views over those advocated in any other work known to me." The immediate benefit of Lyell's book was to hone Darwin's skills as a field geologist. When he returned to England he used Lyell's methodology of interpreting the past by reference to the present, to unravel the problem of the origin of species: "It appeared to me that by following the example of Lyell in Geology, and by collecting all facts which bore in any way on the variation of animals and plants under domestication and nature, some light might perhaps be thrown on the whole subject."

Darwin, like so many others, was in no doubt of Lyell's enormous contribution to geology. But although *Principles* was generally well received, parts of the work never gained broad acceptance. Indeed, one part of his thesis was so preposterous that he was roundly lampooned by De la Beche in one of his most celebrated satirical cartoons. Before looking at this negative aspect, it is important to consider the broader picture.

By the time *Principles* was published most intellectuals agreed with Lyell's position that the Genesis account of the early history of the Earth was untenable. Lyell rejected Cuvier's catastrophic explanation for the geological record and, by extension, Buckland's diluvial deliberations. As far as Lyell was concerned, one of the greatest impediments to progress in geology was the failure to recognize the extensive duration of the geological time scale. Thinking in terms of millennia rather than millions of years had necessarily telescoped geological events, giving an erroneously allegro rhythm to the tempo of change. Lyell frequently used analogies to get his points across and illustrated the time-scale problem by reference to the Great Pyramid at Giza. He argued that if we believed it had been built in only a day, we could justifiably attribute the feat to superhuman powers. Similarly, if we thought a mountain chain had been formed in just a few years it would seemingly require more violent movements of the Earth's crust than presently occur. Lyell's point was that if we took proper account of the vastness of geologic time, the physical changes that have sculpted the Earth could be accounted for by modern causes, acting at similar intensities as they do today. There was therefore no need to invoke any extraordinary, catastrophic explanations to account for geological changes of the past.

Lyell emphasized the nondirectional nature of geological change. At any particular time the land in a given area might be elevated by seismic activity, only to be depressed at some later period. His point was eloquently made by reference to the Roman "Temple of Serapis," near Naples, on the Mediterranean coast. Significantly, he used a drawing of the ruins as the frontispiece of the first volume of *Principles.* Incidentally, the ruins are now recognized as the portals of a Roman marketplace, not a temple to Serapis, the Egyptian deity that the Romans adopted.

The columns of the "Temple of Serapis."

Lyell visited the ruins in 1828, and his investigations enabled him to draw some remarkable conclusions about the building's ancient past. Only three columns were still standing, each forty-two feet high. Until their excavation in 1750, the lower twelve feet or so of the columns were buried in tuff (consolidated volcanic ash). Consequently, the lower segments looked fairly pristine. But above this were segments of similar extent that appeared to be corroded and encrusted. Lyell attributed this appearance to boring organisms, primarily to a marine bivalve named *Lithodomus,* whose remains could still be found in some of its burrows in the columns.

Lyell's drawing of the ruins shows the water reaching to the base of the columns, and the marine borings extending some twenty-three feet above this level. Clearly, the building was once partly submerged in seawater, during which time the boring molluscs attacked the parts of the columns not protected by the embedding volcanic ash. Since that time the sea had receded, leaving most of the columns stranded above the high-water mark. Lyell rejected the idea that sea levels had fallen since Roman times because the remnants of ancient docks, which were dotted around the Mediterranean, were still at the water's edge. Given that sea levels had not fallen, the land must have been raised.

The "temple" was built in the third century, and Lyell reasoned that the Romans had obviously not constructed it with its pillars standing in the sea. The land must therefore have subsided before the building was raised. When Lyell looked into the history of the area he found that a nearby volcano, Solfatara, had erupted in 1198. This must have been the source of the ash, and he reasoned that the earthquakes that preceded the eruption caused the land to subside. A major earthquake occurred in 1538, resulting in the creation of another volcano, aptly named Monte Nuovo. This event would have accounted for the elevation of the building. As eloquently discussed by Stephen Jay Gould, the story is actually a little more complex than this, and was elaborated upon by Lyell in later editions of *Principles.*

This graphic example of how massive changes in the Earth's crust could be caused by processes observable today—modern causes—won Lyell many converts. Few of his readers would have doubted the cyclic nature of geological change. However, his argument that the fossil record was similarly nondirectional met with considerable opposition.

Lyell acknowledged that the idea of a "progressive development of organic life, from the simplest to the most complicated forms" was well entrenched in peoples' minds. However, he claimed it was easy to show that this had no foundation in fact. As an example of the "fallacy," he pointed to the supposed late appearance of mammals and birds in the fossil record. Given that the earliest and most abundant fossils were marine invertebrates, Lyell asked why one would expect to find the remains of land animals at that low level too. He compared the likelihood with that of a sailor dredging up the remains of leopards and elephants from a coral seabed in the Indian Ocean. Just because birds and mammals were not found in the lowest strata did not mean they did not occur at that time too. To add credibility to his claim for the early appearance of mammals, he drew attention to the discovery of the Stonesfield "opossum" in strata that were contemporaneous with the ancient giant reptiles. As far as Lyell was concerned, then, the progression of life through geological time was an artifact of the fossil record, and there was no trend, no evolution, of living things, from the primitive to the more derived states.

Resurrecting an idea dating back to antiquity, Lyell argued that the Earth's climate went through cyclic oscillations about a mean. There had therefore been an alternation between cold periods and warm periods throughout geological time, just as the weather cycles between summer and winter throughout the year. These changes were caused by cyclic changes in the Earth's crust. Given that each species is adapted to a particular climate—elephants to Africa and penguins to Antarctica—Lyell reasoned that this periodicity in climate would result in a cyclic appearance of species through time. This led him to the startling conjecture that a time might come when the "huge *Iguanodon* might reappear in the woods, and the ichthyosaurs in the sea, while pterodactlye might flit again through umbrageous groves of tree-ferns."

De la Beche probably reached for his sketchbook after reading this particular passage, perhaps with an evil glint in his eye. The cartoon he penned is titled, "Awful Changes: Man only found in fossil state—Reappearance of Ichthyosauri." He depicts an ichthyosaurian professor addressing a toothy class of saurians. The subject of the ichthyosaur's lecture is a human skull, propped up against a lithic lectern. The ichthyosaur tells his class that the skull obviously belonged to one of the "lower order of animals," and he wondered how the creature could have procured food with such insignificant teeth and trifling jaws.

Although most of the first volume of *Principles* was taken up with geology and the Earth's changing climate, most of the second volume was devoted to animals and plants. Lyell disagreed with Cuvier on the question of catastrophes, but he agreed with him on the reality, and permanence, of species. Recall that "permanence" was understood to mean that species did not change (evolve) through time, not that they escaped extinction. Such permanence was contrary to Lamarck's thesis of the transmutation of species, which Lyell discussed in some depth. He admitted that transmutation had the advantage of dispensing with repeated acts of creation, but he challenged Lamarck to give concrete examples. Some of Lyell's arguments with Lamarck centered on the changes that had occurred during domestication. If, as Lamarck contended, domestic dogs were transmutated wolves, why, Lyell asked, did they not revert to wolves when they escaped back to the wild? On a more fundamental level, if living things continuously transmutated into more complex ones, why did simple ones, like the microscopic organisms found in a droplet of water, still persist?

Lamarck's response to this question, Lyell informed his readers, was that these "simple" organisms were repeatedly formed in nature, as the raw materials of organic change. But Lyell was not persuaded by Lamarck's transmutational arguments. Besides, the proof that species did not undergo transmutation over time had already been provided by the mummified animals discovered in Egyptian tombs. As noted before, these animals were no different from living ones. Lamarck's argument that this was because there had been no significant environmental changes in Egypt since the time of the pharaohs did not impress Lyell, because the embalmed animals were no different from those living outside Egypt either, where the environment *was* different.

Without attempting to explain how new species may have come into being—a task tackled by Darwin almost three decades later—Lyell accepted that new species *had* been brought into existence. He then proposed that new species arose at "*centres* or *foci* of creation . . . as if there were favourite points where the creative energy has been in greater action than in others. . . ." (Lyell, like most of his contemporaries, used the term "creation" here in a general way, without any implication of divine intervention by a Creator.) Lyell's words have a familiar ring, reflecting our current understanding that speciation probably takes place in small, isolated populations. Darwin expressed a similar sentiment after visiting the Galapagos Islands in 1835. Witnessing the great diversity of species, he acknowledged his astonishment "at the amount of creative force, if such expression may be used, displayed on these small, barren, and rocky islands. . . ."

To summarize Lyell's reading of the rocks, the Earth had undergone a series of cyclical changes during its long history. These nondirectional changes were no different in kind, or magnitude, to those that affect the modern world. The climate had similarly undergone cyclic changes. So too had the plants and animals, each one adapted to a particular clime. As new species arose, other species became extinct. Catastrophism, as championed by Cuvier, had no more place in Lyell's world than did Buckland's diluvial interpretations. Lyell completely rejected Lamarck's speculations on species transformation, along with the widely held view that the fossil record was manifestly directional.

Mantell's prediction of the saints being in uproar probably overstated the public's reaction to *Principles* on its publication in the early 1830s. This is largely because so much of what Lyell presented had already been said before, mostly by Hutton during the previous century. However, nobody had ever marshaled such an impressive weight of evidence to support uniformitarianism, and this must have dragooned many deserters from the catastrophist camp. But

Lyell's unorthodox view that the fossil record was nondirectional won little support. His minority position would become even more untenable as more fossils were discovered, filling the gaping holes in the fossil record and giving further support for its directionality. Surprisingly, though, a similar view would be expressed by another influential man a decade later, the man who coined the name "dinosaur."

It has been said that Lyell was to geology what Darwin was to biology. This does not overstate the importance of *Principles,* nor of its author. It should also be remembered that Lyell did not confine himself to rocks, having devoted most of his second volume to living organisms. His secular treatment of species was especially useful here because it engendered further discussion of what Darwin called "that great fact—that mystery of mysteries—the first appearance of new beings on this earth."

Like Darwin, Lyell was an astute observer as well as a persuasive writer, and much of what he wrote is still valid today. Indeed, it is still fruitful to interpret the past by reference to the present. This has not materially changed with the recent revival of interest in catastrophism, occasioned by recognition of the impact event that occurred at the end of the Cretaceous period. The extraterrestrial body that collided with the Earth 65 million years ago, seemingly destroying the dinosaurs, has all the appearance of a catastrophic event. However, it appears that such events may occur periodically, with a weak frequency of about 40 million years. Accordingly, if we follow Lyell's example and take account of the enormity of the cosmic calendar, these catastrophic events have a regular periodicity, and in this regard could be considered "ordinary" causes.

Buckland may have found Lyell's evidence for a noncatastrophic interpretation of the geological record persuasive, but not compelling. His mind was already made up, and it would take a great deal more than the publication of *Principles* to change it.

9

More Giants

The one certainty about fossil hunting is its uncertainty. Mantell continued to find isolated teeth and scattered bones of *Iguanodon* for more than a decade after his initial discovery, but there was no sign of any associations of bones and teeth, far less a complete skeleton. Nor were there clues that any new giants were lurking in the rocks. Contrast this lack of success with our modern times, where hardly a year goes by without the announcement of yet another spectacular dinosaur discovery. And the rate of discovery has been accelerating during the last few decades. Up until 1969, 135 years after Buckland's discovery of the world's first dinosaur, the number of valid dinosaur genera was only 170. Twenty-five years later this number had jumped to about 315, an increase of some 85 percent. This is largely because vast areas of rock exposures are being searched, for the first time, sometimes in new localities like South America and Africa. Mantell, in contrast, was restricted to making the rounds of a fairly limited number of active quarries in his neighborhood. Like a gull following a fishing boat, he had to be content with any scraps that might be tossed his way.

Mantell may have wondered whether he would ever have another important find in the rest of his geological career. Then something interesting turned up: a new kind of giant reptile entirely different from his now famous *Iguanodon*.

During the summer of 1832, Mantell was on his way to visit the Tilgate Forest quarry when he noticed some large rocks that had been thrown down by the roadside. What attracted his attention was that they contained traces of fossil bone. On reaching the quarry he found that the workmen had put some similar pieces of rock aside for his inspection. They had once been part of a

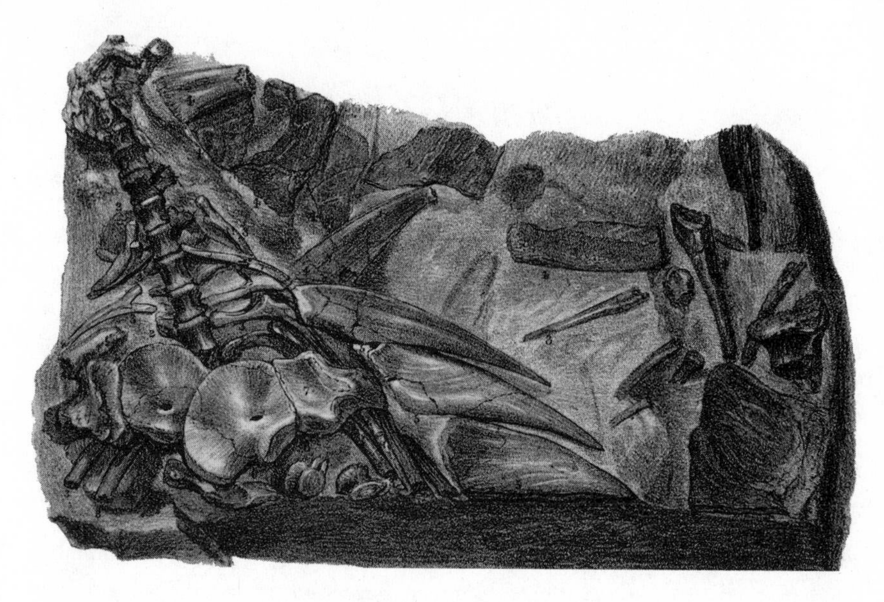

Mantell's illustration of *Hylaeosaurus,* in the block in which it was discovered.

large block that contained the earthly remains of a giant lizard. Unfortunately, the block had been smashed to pieces before this had been realized. The fragmented parts now lay in scattered disarray.

The case seemed utterly hopeless. The few bone fragments visible on the surface looked thoroughly unpromising, and the rock was exceedingly hard, which would make preparation all the more difficult. But Mantell refused to be deterred. He collected every last scrap of rock showing evidence of bone, and had it sent back to Castle Place. After many hours of patient effort, he succeeded in cementing the block together again. It was over four feet long and three feet wide.

He then spent every spare moment for the next several months chipping away at the recalcitrant rock, until he had uncovered all of the hidden bones.

Unfortunately, only a part of the front end of the animal was preserved, comprising a string of a dozen or so neck vertebrae, many ribs, and the shoulder girdle. He named the new fossil *Hylaeosaurus,* meaning "woodland reptile," as it was discovered in the Tilgate Forest assemblage.

The most singular feature of *Hylaeosaurus,* setting it apart from *Iguanodon* and all other extinct reptiles then known, was a series of bony plates and spines. The largest spines were seventeen inches (42.5 centimeters) long and over seven inches (17.5 centimeters) wide at their base. There was only one set of plates and spines, and these lay along the animal's left side. Mantell concluded that, in life, they ran along the top of the reptile's neck and back, like the dorsal frill of many modern lizards: they must have become pushed over to one side during preservation.

The great anatomist Richard Owen, writing about *Hylaeosaurus* several years later, acknowledged that Mantell's "ingenious suggestion" regarding the anatomical arrangements of the plates and spines had initially convinced him that Mantell had correctly identified these unusual structures. However, after further study, he concluded that they were probably abdominal ribs instead. These structures, which are not true ribs, are found embedded in the abdominal region of many ancient reptiles, including ichthyosaurs and plesiosaur. They are also found in modern crocodiles.

As it happens Mantell and Owen were both wrong. The structures were not abdominal ribs, nor did they run along the reptile's back. Instead, the spines and plates projected from the sides of the body and were part of the beast's defensive armor. *Hylaeosaurus,* which was about thirteen feet long (4 meters), is recognized today as an armored dinosaur—a nodosaurid ankylosaur.

As incomplete as it was, this remarkable specimen established the existence of a third kind of giant land reptile, distinctly different from *Iguanodon* and *Megalosaurus.*

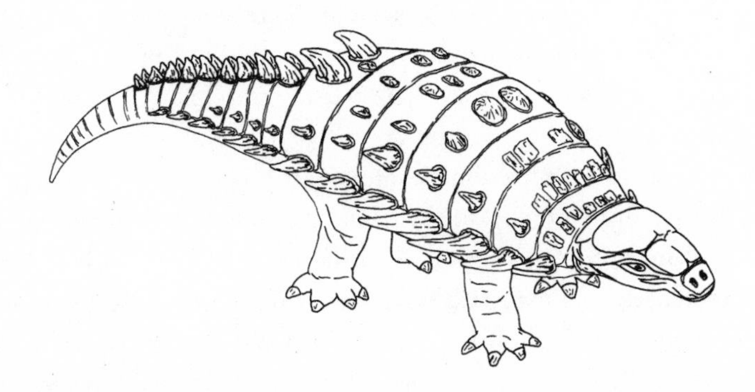

A tentative restoration of *Hylaeosaurus.* We cannot be sure how the animal looked in life because we have so little material. Nodosaurid ankylosaurs are those lacking tail clubs.

Two years after discovering *Hylaeosaurus,* Mantell, now living in Brighton, received some exciting news from Kent, the adjoining county to the north of Sussex. Some workmen at a quarry near Maidstone had been blasting a large block of limestone. When they picked up the pieces after the explosion, they noticed that some had fragments of brown material embedded in them. Thinking this was petrified wood, they showed some of the pieces to the quarry owner, W. H. Bensted. He immediately recognized that the material was fossil bone, not wood—his men had found the remains of some gigantic animal. He directed them to collect every last piece. Once that was done, he set himself to the task of cementing all the pieces together again, as Mantell had done for *Hylaeosaurus.*

When he had reassembled all the pieces of rock containing bone, he had a block of stone that was approximately six feet cubed. Bensted, skilled with a hammer and chisel, then assiduously cleared away the investing limestone, as far as the brittle condition of the bone would allow.

By the time Mantell arrived in Maidstone, Bensted had exposed enough of the specimen to show its resemblance to *Iguanodon.* But there was still a great deal more preparation to be done. Mantell offered him £10 for the specimen, but he declined—he had already been offered £20 but wanted £25. Mantell returned to Brighton bitterly disappointed at losing such a magnificent specimen.

Several weeks passed and Mantell tried to put the whole sorry matter out of his mind, when, out of the blue, Bensted called at his house to deliver the specimen. Unbeknown to Mantell, some of his friends in Brighton had got together and purchased the specimen on his behalf. He was as overwhelmed by their generosity as he was with the acquisition of such a promising specimen. And "now for three months hard work at night with my chisel . . . ," Mantell wrote in his journal.

As it happened, it took him only a month to prepare the specimen. The jumble of bones, all belonging to the same individual, included a virtually complete hind limb, the entire pelvic girdle, parts of the shoulder girdle, numerous vertebrae, and many rib fragments. And, nestled away within this osseous cornucopia was a tooth—a tooth that unmistakably belonged to *Iguanodon.* Here, for the first time, was an associated specimen of *Iguanodon*—all the bones belonging to the same individual. Because the bones were associated with an identified tooth there could be no doubt that they belonged to *Iguanodon.* Many isolated bones similar to these had been found in the past and tentatively identified as *Iguanodon*. Here now was proof that those identifications had been correct. The importance of this specimen was that it allowed similar parts of the skeleton of other specimens to be positively identified as *Iguanodon*, without any reference to the teeth. It also provided valuable new information on the anatomy of *Iguanodon*. But there was still a great deal more to be learned. Nothing was known of the skull, nor of most of the anterior part of the skeleton. This was the most complete exemplar of the genus, however, and it took pride of place in Mantell's private museum.

Interest in *Iguanodon* and the prehistoric times in which it lived was by no means confined to geological circles. There was considerable public interest in geology and paleontology during the 1830s. Mantell had done much to popularize his subject and had become quite well known in the process. This was partly through his museum with its celebrated collection, and through his writing. Aside from writing books he was a regular contributor to the newspapers and wrote a regular column, "News in Science," for the local paper. But it was mostly through his public lectures that he had become such a well-known local figure. These lectures were hugely successful, with audiences upward of 800. They were widely and favorably reported in the press too. Part of their success was because they were well advertised, through handbills,

The *Iguanodon* quarry at Maidstone, Kent.

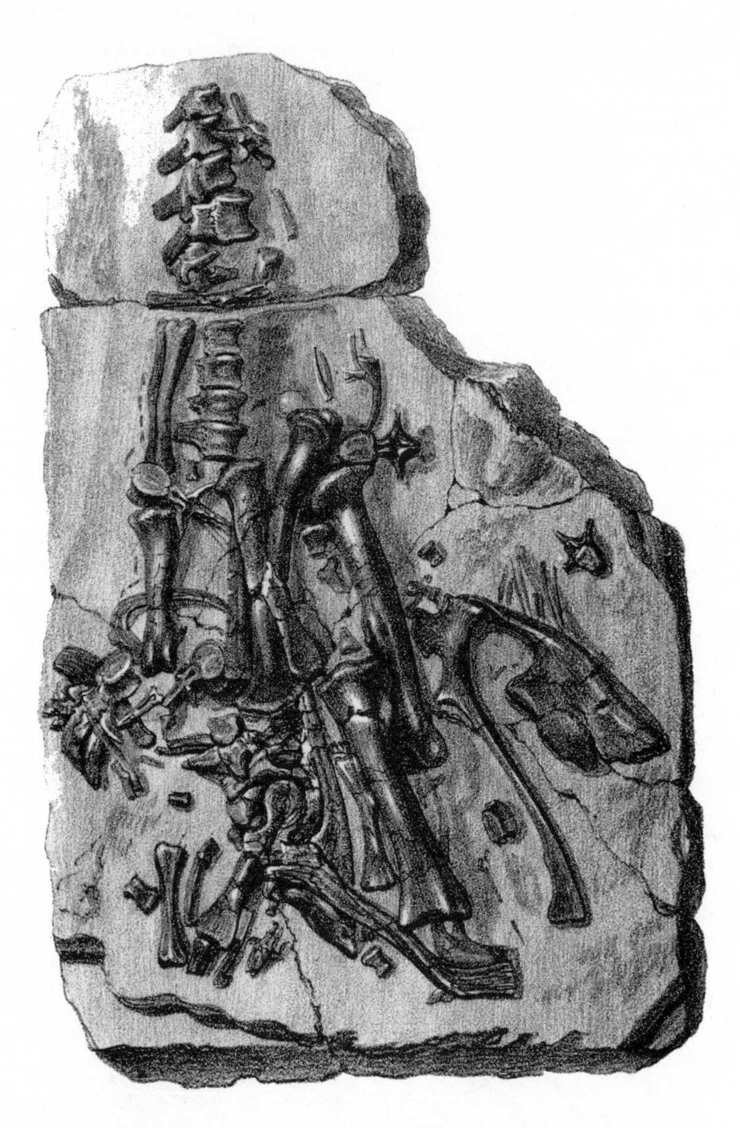

The Maidstone *Iguanodon,* as illustrated by Mantell, after he had chiseled away the overlying rock from the bone.

posters, and advertisements in the local press: "On Saturday, 8th February, 1834, A Lecture on Geology and on the Organic Remains of a Former World, Discovered in Sussex, will be delivered at the Old Ship, at Two o'Clock, by Gideon Mantell, Esq. F.R.S. &c. (Late of Castle Place, Lewes.)."

Tickets for this particular lecture were two shillings and sixpence [one-quarter of a pound], a large sum of money back then. Mantell had been living in Brighton for less than two months, and he therefore began his lecture by assuring the resident physicians that he had not come as a rival. He wished to settle among them on equal terms, and when he announced that the proceeds from his lecture were to be donated to the Sussex County Hospital, there was an enthusiastic round of applause. He also sought to assure potential patients that his interests in geology would in no way detract from his attention to their medical care. Having dispensed with these professional matters he treated his audience to an exposition of prehistoric life in Sussex, with emphasis on the great reptiles that roamed the land. "One of the most interesting portions of the lecture," the *Brighton Guardian* reported afterward, "was the juxta-position of the thigh bones of the small Iguana, and of the mighty Iguanodon. . . . [His researches] enabled him to proclaim to the astonished scavans [*sic*] that there existed formerly in England a reptile 100 feet in length!"

We know today that *Iguanodon* was only about thirty-three feet long (10 meters). Mantell arrived at his exaggerated size estimate by scaling up an iguana lizard until its femur was the same length as that of *Iguanodon*. This reflected the widely held view of those times that *Iguanodon* and its reptilian kin were merely scaled-up lizards. Even if this were true, it would still have been inappropriate to use a simple one-to-one scaling of femur length to estimate body length. This is because the various parts of an animal's body do not usually increase in size in simple step with body length. The most familiar example of this is the growth of the head. Immature individuals, whether puppies or human babies, have relatively huge heads compared with their bodies. But their heads do not continue to grow as fast as their bodies, so when they attain adult size their heads are relatively small compared to the rest.

The popularity of Mantell's talks was largely owing to his lecturing skills, but it also reflected the public's voracious appetite for knowledge of science and technology during those rapidly changing times. People from all levels of society clamored for knowledge of the new age, attending lectures on steam locomotion or electricity as we might attend the cinema. Their craving for knowledge was fed by public lectures, demonstrations, exhibitions, and by

Mantell's vision of *Iguanodon* as a scaled-up lizard. Modern restorations are drastically different, and depict a bipedal posture.

published articles. These appeared in the daily press and in a variety of magazines and journals, from popular publications like the *Saturday Review* to more exclusive periodicals like the *Gentleman's Magazine,* the *Magazine of Natural History*, and the specialized journals of numerous associations and learned societies.

Seeing the popularity of his public appearances made Mantell wonder whether lecturing might provide him with an alternate source of income. And Mantell desperately needed money: His move to Brighton had been a complete financial disaster. Visitors packed into his museum and large crowds attended his public lectures, but patients stayed away from his surgery in droves, believing that his devotion to geology would detract from his medical practice. Thinking his lecturing could resolve his financial problems, he brushed up on his comparative anatomy, hoping to broaden his audience with a series of lectures on a medical theme. The lectures were a great success and were very well received, as reported in the *Lancet:*

> The course was illustrated by diagrams . . . which, although strictly and anatomically correct, were deprived of every repulsive character. . . . Thus, objects which are generally considered unattractive, and even disgusting, were rendered very engaging by the skill and tact with which they were discussed: and ladies of rank and fashion were seen handing round glasses containing dissections of the eyes. . . .

In spite of good attendance at his lectures, it soon became apparent that they would provide little more income than his spluttering medical practice. He contemplated seeking his fortunes in the United States, and wrote to seek the advice of his friend, Professor Benjamin Silliman of Yale University. The two men had corresponded with one another for many years, and had formed a close friendship, though they never did meet. Silliman consulted several of his own friends, but they were all of the opinion that Mantell would encounter the same misgivings in the New World over the perceived conflict between geology and medicine as he had in England. None of this helped Mantell's melancholic state of mind.

His financial worries were partially relieved when he entered into a business arrangement with some of his Brighton associates in the spring of 1836. Under this arrangement his private museum collection, and the house in which it stood, became the Sussex Institute and Mantellian Museum. In consideration for the loan of his specimens and his home, he received an annual payment of £350. This, of course, meant vacating the house. He moved into lodgings, just along the road from the newly formed institute. Meanwhile, his wife and family moved back to Lewes, where they lived in a small rented cottage. Their youngest daughter, thirteen-year-old Hannah, moved to Dulwich, on the outskirts of London, where she attended school. The reason for the breakup of the family, which was not intended to be permanent, was partly fiscal and partly domestic. Things between Mantell and his wife had not been as they should for some time. His obsession with fossils and his merciless driving of himself in the unending quest for the unobtainable had taken its toll on the marriage. The decision for the family to return to Lewes was probably made because they were happier there. If the truth be known, he might have been happier there too. But he could not cut his losses and return to medicine in Lewes because the man who had purchased his practice was doing rather well and there was no room for another physician in that small town. He must try to weather the storm.

Although he lived apart from his wife and family, relations appear to have been cordial, and there were periodic visits. For all that, Mantell was sad and lonely. To add to his depression, Hannah's health seemed to be failing. She would eventually die four years later, devastating her father. There is a poignant journal entry for May 19, 1836:

> Very miserable as usual! found on my table a beautiful Vase with flowers and this billet—
>
> Quand la beauté de ces fleurs sera flétrie
> Que l'amie qui les donne soit encore chérie
> Elle ne t'effacera jamais de sa memoire
> Dans tes pensées qu'elle ait un séjour!
> [When the beauty of these flowers will be faded
> That the love that they gave would be still cherished
> The beauty will never be erased from your memory
> In your thoughts it would have a dwelling!]

I believe this was penned by his wife because two days later he went to Lewes and she returned with him. She subsequently took up residence in Brighton with her two daughters.

The uncertainties in Mantell's life continued beneath cloudy skies, with only the occasional patch of blue. At times he appeared to be living with his wife and family, and then to be on his own again. The health of fourteen-year-old Hannah remained of great concern to him, and his relationship with Ellen, the eldest, deteriorated. She was now a young woman of eighteen and appeared to resent her father for what he was doing to her mother and to their family. The oldest son, Walter, now sixteen, was following in his father's footsteps and had started a three-year apprenticeship with a surgeon in Chichester. Reginald, the youngest, who was only nine, attended school in Brighton. The settlement for Walter's indenture with the surgeon, and the continuation of school fees for the two youngest children, were a tremendous drain on his resources. His financial prospects were now so unsure that he realized he would probably have to sell his fossil collection, still on loan to the Mantellian Museum. But before he lost control of his own fossils, Mantell would become involved in deciding the fate of another private collection. This particular collection was as remarkable for its eccentric owner as for the exquisite skeletons it comprised.

10

Improving on Nature

Thomas Hawkins was an odd character. A contemporary described him as being of middle height, with fair hair and a foxy, unpleasant face. When Owen first met him he described him as a "worthy and eccentric man of genius." Mantell said the twenty-two-year-old was "very romantic, very weak, very good natured, & I fear very headstrong." The Somerset locals of his Glastonbury hometown, near Street, may have viewed Hawkins as a gentleman of means, with nothing better to do with his time or wealth than collect a lot of old fossils that were of no use to anybody. But just how wealthy he was is difficult to say. It has been reported that he had a modest inheritance of £200, with a £10 annual rent from the family orchard—no fortune even in those times. Mantell recorded that he had been left a "few thousands," but I suspect this may have been the result of his bragging to impress Mantell, and that he was probably of limited means. His father had died around 1830, when Hawkins was twenty. Hawkins senior was a farmer and cattle dealer, and had encouraged his son's early geological interests by giving him a sufficiently generous allowance to purchase fossils: "my ardour and liberal allowance of money secured me a very fine collection before I numbered twenty summers. . . ."

The part of the county where Hawkins lived lies on the same band of lias that outcrops at Lyme Regis, though the Somerset rocks are geologically slightly older, and the fossil fauna marginally different. A number of small quarries once dotted the countryside, but they have long since disappeared: overgrown or built upon. A few remnants can still be seen, however, if one knows where to look. The quarries were worked for their limestone, which had a variety of uses, from building stone and road metal to making lime for cement. Pieces of the liassic masonry can still be seen in the ruins of Glastonbury Abbey, overgrown by ivy and foxgloves, and elsewhere. Sometimes the quarrymen found fossil bones embedded in the limestone, which they believed were "the bones of infants and giants."

As a youth Hawkins made regular visits to the quarries in and around the villages of Street and Walton. He chatted to the quarrymen to see what fossils they had found, and may have done some digging himself. But it was an uphill battle for him at first because most of the quarrymen, owners and laborers alike, would just as soon put their hammer to a fossil as save it. Hawkins recorded one such incident in which a quarryman was about to destroy the lower jaw of a plesiosaur. Hawkins's timely intervention saved the specimen. By paying the quarrymen generously for the trifle he made sure that future finds would be saved.

Although most of the quarrymen had no interest in fossils, there was a notable exception in one George Moon, a laborer at a quarry in the village of Walton. He used to collect the fossils that turned up in the quarry, though he made no attempt to disabuse his workmates of their naive beliefs in them. Nor is he likely to have gone to great lengths to save a fossil from destruction, at least not until he met Hawkins and discovered they had some monetary value.

Hawkins was so passionate about geology that he enthused everyone he talked with. One warm July evening in 1830, he called on his old schoolteacher who lived in a hillside cottage overlooking Glastonbury, the legendary burial place of Arthur and Guinevere. Hawkins told him of the wonders of geology, and, elevated by his story, they set off together in search of fossils. They walked unhurriedly down country lanes, enjoying the scents and sounds of summer: honeysuckle and droning bees, fresh hay and the bleat of sheep. Whichever way they looked brought another magnificent scene into view. But the dominant feature, across fields and moors and low rolling hills, was Glastonbury Tor. Rising like a mountain peak, the 500-foot hill, topped by its ancient tower, had been a landmark around these parts since time began, for all anyone knew. And away off in the distance, burnished by the lowering sun, was the Bristol Channel, with the blue coast of Wales beyond. A milkmaid, heavy with pail, nodded and smiled and passed the time of day.

Coming to the village of Street they made their way down a hedgerowed lane, passing the blacksmith's forge and the tiny cemetery. They stopped outside George Moon's whitewashed cottage, where his wife told them her husband was still at his work. She then produced part of the tail of an ichthyosaur, discovered the previous day.

"I must see Moon, my good woman," exclaimed Hawkins excitedly, and set off for the quarry at a pace, his old schoolteacher in tow.

They met Moon and his fellow quarrymen along the road leading to the quarry. An animated Hawkins asked Moon to take him back to the excavation and show him where the specimen had been found.

Arriving back at the quarry, Moon pointed down at the bottom of the freshly cut face, about twelve feet below ground level. Hawkins saw the tell-tale bone peeking from the rock. He was down there in a trice, scrambling in the dust on his hands and knees. He identified the protruding objects as a section of a rib and part of a vertebra.

"We must dig it out tonight, hearties!" exclaimed an enthusiastic Hawkins. "We can't, zir," said Moon, "'tis too leate." But Hawkins was insistent: "We will!"

There probably followed a brief and private exchange between the two men, where a price was agreed upon for the fossil. Hawkins always gave "the master one half, and his man who chances to find it the other." Sleeves were then rolled, tools taken up, and the sound of hammers set roosting birds to flight.

Night fell and candles were lit. The retired schoolmaster took his leave at that point and set off for home, but Hawkins remained at the quarry, his enthusiasm undiminished by the lateness of the hour. Hammers rang out into the darkness.

Eventually the workmen, exhausted by their long day's work, had to put down their tools. "George," Hawkins told Moon, "you will stay here all night to see no mischief is done . . . ," for he feared "the rude curiosity of the simple villagers." He then walked home to Glastonbury in the dark, his mind perhaps filled with dark images of monstrous creatures, locked in mortal conflict.

Hawkins had a fire-and-brimstone image of the Earth during the days of the saurians—that remote period of time before God created Adam and Eve and all the familiar creatures that inhabited the modern world. For him the globe sweltered beneath a sunless firmament, a world of perpetual darkness and loathsome struggle, that was a "fitting habitation for the cold-blooded reptiles" that dominated the Earth. He commissioned the artist, John Martin, to engrave such a scene for the frontispiece of his second literary work, *The Book of the Great Sea-Dragons,* published in 1840. Two long-necked plesiosaurs are depicted, just offshore, attacking a third creature that could be an ichthyosaur. The beached carcass of another ichthyosaur is shown in the foreground, being scavenged by a gaggle of leather-winged pterosaurs. One of their number picks away at the dead creature's eye. And in the background, barely visible in the deep gloom, are the coils of some unidentified sea monster. This was the stuff of nightmares, but Hawkins was in no doubt of the malevolence of that dark age before ages.

The next morning found Hawkins, still in his dressing gown, hammer and chisel in hand. Moon and his men had delivered the specimen sometime after dawn, and Hawkins was hard at work chipping away at the rock that covered the skeleton. The zealous fossilist had already uncovered part of the tail and snout. Most of the rest of the skeleton still lay beneath a thin layer of rock, discernible by the slight undulations over the underlying bones—the meat beneath the pastry. The skeleton had probably been removed from the quarry in a number of slabs, and these might now be arranged together on a stout table, in their correct order, like a paving-stone jigsaw puzzle.

Hawkins was a flamboyant character, an eccentric who held fixed views, with the firm conviction that he was always right. He had no doubt that the sea dragons of the ancient Earth were extinct: "They have ceased from off the face of the Earth: inexorable time long since extinguished the last of their race and all that survives of these once-grim and omnipotent aborigines are a few crushed bones as unsightly as they are rare."

The antediluvian world of the saurian was as remote and alien to Hawkins as the moon, and he could never concede how past geological events could possibly be explained in terms of present-day changes. He would therefore never be reconciled to Lyell's principle of interpreting the past by reference to the present. As far as the relationships among species were concerned, Hawkins acknowledged that there were similarities between some types and others, with close gradations in many instances—the links in Buckland's chain of living beings. But he did not believe this could be taken as evidence for the transmutation of species, "a doctrine as unphilosophical as it is impious," and he railed against "the Lamarkian [*sic*] heresy." Instead, Hawkins viewed the gradations between species as evidence of "the consummate skill of the ever-to-be-worshipped Creator's purpose."

Hawkins was extremely skilled in the art of fossil preparation, doing most of his work with a hammer and chisel. It took him a month's hard work to remove the overlying rock from the seven-foot ichthyosaur from Moon's quarry. But fossil preparation involves more than just the removal of unwanted matrix from a fossil. It also includes strengthening exposed bones and repairing broken ones. If the specimen is primarily intended for display purposes, it may also involve the restoration of missing parts.

Hand preparation is still used today, but we also have a variety of mechanical devices to do the job. These include the "Airscribe," a handheld pneumatic chisel driven by compressed air, and the air-abrasive, a device that delivers a fine stream of high-speed abrasive powder such as carborundum through a pen-sized hand tool. Rocks containing limestone can also be removed chemically by immersion in dilute acetic acid. This dissolves away the matrix without affecting the fossil bone. Acid preparation works extremely well upon the ichthyosaur and plesiosaur material from the southwest of England, but it was not introduced until the 1950s.

The usual way of preparing an ichthyosaur or plesiosaur for display, today as in Hawkins's time, is to start by removing the overlying rock from the surface of the bones. Missing bones, or parts of bones, are often replicated in plaster, or in polyester resin, by reference to other specimens in which the feature is intact. Sometimes the restored parts are realistically colored, to blend in with the rest of the bone, but this is not recommended because of the confusion it causes researchers. For example, if a paleontologist were measuring the dimensions of a bone that had been seamlessly restored, the data would have no validity. Similar errors would arise if parts of a skeleton had been restored by the addition of missing bones, either casts or originals, from another individual. It is therefore important to distinguish the restored bones by coloring them differently.

Since most skeletons are brought back from the field in several slabs, the last step in preparation is to consolidate the slabs. This was often done, as we have seen, by building a wooden frame around the laid-out skeleton and filling all the gaps with plaster to form a wall-mount. Most ichthyosaurs and plesiosaurs have been flattened to a thickness of only a few inches by the pressure of the sediments that became compressed into rocks. This is why they are left in the rock and displayed as wall mounts. Very occasionally, though, they are found in a relatively uncrushed state, like most of the dinosaur skeletons we see in museums. In these instances the individual bones can be completely removed from the rock and mounted as a three-dimensional skeleton.

Hawkins probably repaired broken bones with animal glue, often strengthening the bones and the surrounding matrix with acacia gum. Today we have a wide choice of synthetic adhesives for repairs, including acrylics, cyanoacrylates ("Krazy glue"), and epoxy cements, together with a variety of polymers for strengthening. He would have restored missing parts with bones from other specimens or by fabricating them in plaster. The fabrications were then realistically colored to match the rest of the skeleton, his primary interest, as we will see later, being to restore the skeleton to a state of near perfection.

Hawkins did most of his collecting in and around Street, but he also visited other localities, including Whitby, in the northern county of Yorkshire, and Charmouth and Lyme Regis in the south. And it was here, in Lyme, that Hawkins undertook the rescue of an ichthyosaur skeleton from the sea. "Every body remembers the wet and wearisome summer of last year [1833]," Hawkins wrote in his first book, *Memoirs of Ichthyosauri and Plesiosauri,* published in 1834, "deluges of rain—deluge upon deluge—floods by land, tempest at sea. . . . " He visited Jonas Wishcombe of Charmouth to see if he had any fossils to sell, and learned he had found an ichthyosaur low on the shore at Lyme Regis. Hawkins wanted to see it immediately but was told that it lay beneath many feet of water. "In marl or stone?" Hawkins asked. "Why in beautiful maarl," replied Jonas, rolling his Rs. "Washed to death," returned a distraught Hawkins, throwing himself in despair upon a chair. Jonas tried to console him, and they agreed to meet at low tide on the morrow to see the beast. Dawn broke and Hawkins prepared for the new day. "The weather-cock looked the right way as we descended the execrable path by which the good people of Lyme are content to wade to the sea-shore. . . . " They reached the spot and Jonas pointed down to the receding waters. "Do you see that tuff o' sea-weed just tipping up there, yer honor?; he's there about. . . . " The tide surged and Jonas got soaked to the knees for his trouble. Hawkins now saw:

> the object of my Egyptian-like idolatry, the adored lizard at my feet. Jonas gladly sells me the right to the skeleton. . . . He chuckled when I gave him a guinea earnest-money, convinced that he had made brave of a discovery that no one could render useful. . . .
>
> "Yer honor sir, I'll be here to-morrow at low tide."
>
> "Aye, and Jonas—bring as many men and tools to help as you think proper."

Their business concluded for the day, the two men made their way back from the beach.

> "You will never get that animal," said Miss Anning, as we made our devious way towards Lyme through the mist and flashing spray, "or if you do, *per-chance*, it cannot be saved."

> My eyes glare upon the intellectual countenance before me—the words of those lips were I knew those of a Pythoness [priestess] and my heart fainted within me.
>
> She saw my change of blood and stopped—"because the marl, full of pyrites [iron mineral], falls to pieces as soon dry."
>
> I revive—"that I can prevent."
>
> "Can you."

Hawkins could scarcely sleep that night as he thought of the sea pounding away at his precious ichthyosaur. "I lay upon a thorny pillow listening the live-long night to the rumbling gale . . . but the day breaks and the wind's the wrong way, south-west."

The worst weather always hits Lyme Regis from the southwest, and the wind continued to blow from that direction for the next three weeks. The wind drove heavy seas against the coast, making the recovery of the specimen impossible. The impetuous young Hawkins must have been at his wit's end. But then he awoke one morning to find that the weathercock was pointing in the opposite direction. He dressed in a hurry and dashed to the beach, passing Mary Anning along the way. They exchanged brief pleasantries and he hurried on, gathering half a dozen lusty workmen to help with the excavation.

By the time they had chiseled out a block of rock containing the skeleton the tide had turned, threatening to inundate the specimen. They tried to raise the heavy block and lift it onto a cart, but their strength failed them. One of the men suggested breaking it in two, but Hawkins would not hear of such a thing and encouraged them to make one last effort. They succeeded, and the "saurus" was saved.

The specimen Hawkins rescued from the sea that summer, over 150 years ago, can still be seen on public display at the Natural History Museum in London. It is a six-foot-long skeleton of *Ichthyosaurus communis.* Hawkins described his preparation of the specimen:

> [It] came forth at the magic touch of my chisel [and] lay like a new creation before me—and I was the creator. I worshipped it for hours in my mad intoxication of spirit. As Miss Anning anticipated, the marl—as soon as it dried—cracked, but by the assistance of some clever carpenters we secured it in a tight case with plaster of Paris so no power can now disturb it.

The ichthyosaur that Hawkins rescued from the sea during the summer of 1833. This skeleton *(Ichthyosaurus communis)* can be seen in London's Natural History Museum.

Hawkins certainly did possess a magic touch when it came to preparing fossils. Unfortunately, the end product often looked more perfect than nature had intended. Mantell, who examined Hawkins's collection after it had been moved to London, recorded: "this gentleman Mr. H. restores them too much: he makes nothing of putting on an arm or a tail, or a rib, where they be wanting: he does not do this without authority; yet still I think it is objectionable when art is allowed to interfere so far."

Hawkins also took great liberties with zoological nomenclature. With total disregard for the scientific names that had already been given to the various species, he published new ones of his own. These names, which have no validity because the original ones have priority, were long and discordant. His replacement name for *Ichthyosaurus communis,* for example, was *Ichthyosaurus chiroparamekostinus,* meaning the ichthyosaur with oblong hand bones.

His replacement names all made reference to the shape of the limbs, which he considered the defining character of ichthyosaurian species. In this, Hawkins was well ahead of his time. A decade before, Conybeare and De la Beche had used tooth shape to distinguish among species. But teeth are of limited value because they are so variable. Fin characteristics, in contrast, can be used to distinguish among certain species.

Reviewers of his book soundly criticized him for his nomenclatural transgressions. Edward Charlesworth, editor of the popular *Magazine of Natural History,* suggested the names should be restricted to the use of Somerset quarrymen, whose West Country accents would produce the most pleasing pronunciations.

Conybeare added a postscript about the book in a letter to Buckland: "What capital fun Hawkins book is. I only wish it had been published before Walter Scot died. It might have furnished him a new character, a Geological bore far more absurd than all his other ones put together."

Aside from these criticisms the book was generally well received, though, like its author, it was an odd mixture. Written in a florid style, it begins with a rambling account of the Creator and of ancient history, underscoring Hawkins's belief in the omnipotence of God as the creator of the world and all its inhabitants, past and present. Most of the rest deals with the saurians and their discovery, with occasional lapses into poetry and rambling passages of purple prose.

The book leaves no doubt of Hawkins's consuming passion for his subject. Nor is there any doubt of his irrationality, foreshadowing the mental instability that would mark his later years. This is more evident in his second book, in which he elevated the ichthyosaurs and plesiosaurs to their own kingdom: "They are the Jews of Creation, alone in the actual World. . . . the Gedolim Taninim, the Great Sea Serpents, the frightful Dragons of the Dead Times, the long lost Ichthyosauri and Plesiosauri. . . . "

His vivid imagination painted the most surrealistic pictures of marine reptiles, like this one of a particular ichthyosaur species: "The Cunning and cruel Snake, whetting his fangs with poison in treacherous lair, and following with malignant eye the unconscious creature of his lust. . . . "

But for me, the most compelling example of his suspension of reality was when he convinced himself that ichthyosaurs had mammary glands. This occurred to him while he was preparing the thoracic region of one particular specimen. He described how his chisel:

> sunk through several thin laminae not unlike charred leaves, which we probed with intense concern. Curious suspicions of Mammae began to haunt us. We had hitherto supposed these Sea-dragons oviparous [egg-laying], and now we are tempted to think of them mammals. . . . If these great Sea-Dragons certainly suckled their impy brood, which these appearances incline us to believe, Martin [the illustrator] has barely

> attained, with all his stupendous Powers, the utter hideousness their own. These huge Dragons and their horrid Brood!

(His use of the royal "we" when referring to himself was not such an unusual style of writing during those times.)

Regardless of these deranged ramblings, Hawkins deserves credit for recognizing that the tail bend—the downward kink of the vertebral column in the tail region—was natural, and not an artifact of preservation as Richard Owen believed: "Nor was the condition of the tail without advantage . . . the declination to which the tail is always found more surely prove that it was naturally so . . . and possessed of a lateral motion peculiarly its own. . . . "

The reason contemporary fossilists were prepared to indulge Hawkins with all his eccentricities, even to admire him, is attributable to the outstanding collection of saurians he had acquired. They were unique and without parallel anywhere in the world. He was a self-taught man, but his lack of any formal geological training would not have weighed against him as heavily then as it would today—many of his contemporaries similarly lacked a formal background. Hawkins apologized to his readers for this deficiency in his second book, saying that he had enjoyed neither the privilege of a mentor nor the means to acquire much worldly wisdom.

Educational shortcomings aside, his lowly birth to a common farmer would have been a serious impediment to his acceptability into society in those rigidly class-conscious times. This helps explain the fawning tone he used when referring to men like Buckland. It may also explain the social airs he adopted later in life, when he proclaimed himself to be the "Rightful Earl of Kent," though this may have been attributable to his unbalanced mental state.

Hawkins was born and raised in Somerset, but from about 1830 until around 1833, when he took up permanent residence at Sharpham Park, near Street, he spent his winters in London. This was primarily to give him access to the important men of science. He was not the least bit bashful about ingratiating himself with influential people, and had a natural ability to charm his way into favor. Buckland, to whom he dedicated the opening part of his first book, was among his first converts.

London was the cultural as well as the administrative capital of the country, with an enormous appeal to residents in spite of its grime. The wealthy could afford to live in secluded luxury, but the working poor, who flocked to the city for employment, were not so fortunate. They lived in close-knit squalor, without adequate water supplies or sanitation. Rat-infested hovels crowded the

Apparently the only existing portrait of Thomas Hawkins, showing him during the latter part of his life.

banks of the Thames and elsewhere, and the river itself was little more than an open sewer. Disease was rife and epidemics spread like wildfire. Hawkins recalled the cholera epidemic of 1831, which caused him to change his plans and leave London for the country earlier than intended. Cholera, introduced into Britain from the Far East, was one of the hidden costs of expanded trade with the rest of the world.

> December [1831] gave up the ghost amidst a thousand frightful rumours of the coming cholera . . . who will ever forget the panic that followed; London was comparatively deserted within twenty-four hours . . . along the Borough bank of the Thames—in those crowded lanes, where so

> many Irish people herd, pent up in a lazar-house . . . what havoc and death! Wednesday fatal cases trebled—about twenty publicly acknowledged—at least a hundred and twenty known to the observant few. Ah! I was smoking cigars on the box of the Bath mail all the night and at ten o'clock Thursday galloping over the Mendips—the British Alps—on the *Exeter*."

Arriving in Street during the first few days of 1832, he made his usual rounds of quarries and discovered that one of the owners, Ephraim Creese, had found a plesiosaur just a few days before. He hurried off to Creese's home to see the specimen, but was dismayed at the amount of damage the quarrymen had inflicted. Having thrown away most of the two front paddles, and all of the right back one, they broke the slab bearing the rest of the skeleton into almost thirty pieces, and stabbed it all over, like matadors.

Creese paid dearly for his carelessness because Hawkins gave him far less money for the specimen than he would otherwise have done. He lectured the quarryman on the need for care in removing fossils, and told him what the intact fossil would have been worth. Nevertheless, he still paid him well for the find—he wanted to keep Creese in his pocket. Hawkins insisted on visiting the quarry, and managed to recover most of the missing paddles by rummaging through the spoil heap. He spent the next two months piecing the skeleton together.

Plesiosaurs, in contrast to ichthyosaurs, were dorsoventrally compressed, meaning they were narrower from top to bottom than from side to side, like a turtle. Consequently, when plesiosaurs died and sank to the sea floor, they nearly always came to rest on their backs or their bellies, and this is the way they are usually preserved. Ichthyosaurs, in contrast, are usually found lying on their sides. Hawkins's fragmented plesiosaur lay on its back, and the specimen can still be seen on display in London's Natural History Museum.

A little under six feet long head to tail, the skeleton lies with its paddles spread-eagled at its sides. The left fore-paddle is incomplete beyond the forearm, but the other three are fairly complete. And the most complete of them all is the right hind paddle, the one the quarrymen had discarded in its entirety. So, unless Hawkins was mistaken about their identities, the most ruined of the four paddles is now the most perfectly restored one. It is difficult to imagine that this specimen was once broken up into dozens of pieces. There are no obvious gaps, joins, breaks, or in-fillings. There are no differences in color or texture between neighboring bones, or between some parts of the matrix and

The near-perfect skeleton of the plesiosaur that Hawkins reconstructed from dozens of broken pieces. It can still be seen in the Natural History Museum in London.

others, and the matrix looks absolutely original. Nor is there any obvious damage to the surface of the bones, though close inspection reveals a few very small pits in some of them. In short, the specimen is nearly perfect. Too perfect.

Hawkins was clearly a master craftsman, an invisible mender who could restore a shattered skeleton to its original splendor. This necessitated the replacement of missing bones and parts of bones. He probably achieved this, where possible, by inserting parts from other specimens. Other missing parts would be fabricated in plaster and color-matched to the original. He was probably also adept at simulating natural rock. Why did he go to such lengths to restore the skeletons in his collection? Was he simply a perfectionist who wanted his specimens to look flawless? Or did he wish to deceive?

Hawkins was not alone in his restoration of incomplete skeletons. Plenty of quarrymen got into the business of collecting fossils, and although it is true that they received more money for a complete than an incomplete skeleton, their motivation to add a missing part from another specimen may not have been purely deceptive. Most of these people had little or no knowledge of the animals they collected, and as far as they were concerned the head of one specimen was just as good as that of another. So what was the harm in completing a skeleton by adding the missing part from another? Perhaps no different from completing a consignment of building stones with stones added from another quarry. However, Hawkins *did* understand fossils, and his intention in restoring his skeletons was likely to deceive—he simply wanted others to think his collection of saurians was the very best in the world.

Hawkins sometimes paid others to assist him with preparation, as when large quantities of plaster had to be mixed and poured, and he sometimes worked on other people's specimens too. In writing to Mantell about the plesiosaur he was wall-mounting for him, he makes reference to the "plaster mans bill." Hawkins is quite clear in his writings that he did the skilled work himself. However, Mantell recorded that he "employed an italian artist to carve them out, & repair them, paying him most liberally: the consequence is that he has the finest specimens of the kind in the world." Regardless of the extent of the skilled labor he employed, Hawkins was ultimately responsible for the authenticity of the end product.

By chance Thomas Hawkins arrived in Lyme on the very July day in 1832 that Mary Anning discovered the skull of a giant ichthyosaur. The monstrous great head was some five feet (124 centimeters) long, making it the largest ichthyosaur that had so far been found. She discovered it at the foot of the cliff, embedded in marl, and probably hired some local laborers to help her dig it out. Soft marl is easily removed but it readily crumbles, and the skull probably broke into several pieces during its removal. The massive ichthyosaur had come to rest on the seabed on its belly, which is atypical, and the pressure of the accumulating sediments had so flattened the skull that the eye sockets were reduced to little more than slits. For all that, it was still a spectacular find, though she could find no trace of the rest of the skeleton. Hawkins purchased the skull, and she agreed to take him to the spot where she had found it.

The two fossilists met the following morning and walked along the beach to the foot of the cliff. When she showed him the excavation in the marl where the skull had been removed he became convinced that the rest of the skeleton was still there, buried in the cliff. Anning was equally convinced she had collected all there was to be found. Hawkins was only twenty-two at the time, eleven years her junior, and with considerably less hands-on experience in collecting fossils. Like the other fossilists, he held her in the highest esteem, but he still thought she was wrong and that the rest of the specimen was in the rocks. Confident of her convictions, she gave him leave to excavate the site. Hawkins obtained permission from the landowner to pull down part of the cliff, and hired men for the job.

Early the next day, before the sun had driven the last of the mist from the hills, they began the task of humbling the cliff. With pick and shovel they toiled, and into the next day, attracting a flock of onlookers from the surrounding countryside. Hawkins had been right, and by noon on the third day they had exposed the entire skeleton. Hawkins recorded how the lusty cheer from the workmen was taken up by the spectators, making "hill and mossy dell echoing ring."

Unfortunately, the skeleton, embedded in the soft marl, broke into myriad fragments. Hawkins despaired of ever reuniting the pieces, even with Anning's kind assistance. But by nightfall the last of the heavy boxes of bones had been packed—the skeleton was safe. Hawkins had the specimen delivered to his summer residence in Glastonbury.

> It arrived at my house the 1st August, at six o'clock in the morning and the whole of the day was occupied in unpacking and disposing in their proper place, as well as I could, the dreadfully numerous pieces. I shall never forget the intense heat of the following fortnight during which time, the Sabbath excepted, I was engaged from day-light to dark-night in developing it. . . . When my manual labours terminated, it counted about six hundred pieces, some of which were so brittle that it was dangerous to touch them. These a trusty Lucchese [a witty, scoffing fellow], under my especial direction, fixed in sulphate of lime [plaster] of which thirty hundred pounds [1.4 tons] were used, in a case that weighed half a ton.

Given that it took Hawkins two months to restore the small plesiosaur he purchased from Ephraim Creese, it seems remarkable that it took him only two weeks to work on this huge skeleton, especially considering his "perfectionist" tendencies.

His reputation for taking liberties with his specimens seems to have been common knowledge among the other fossilists. Mary Anning wrote the following about this particular specimen:

> if Mr. Hawkins has set the last specimen . . . as it lay in the cliff, it will be a most magnificent specimen, but he is such an enthusiast that he makes things as he imagines they ought to be; and not as they are really found . . . I would not have trusted [the specimen] to his making up,

> though very much broken, it might be made a splendid thing without any addition. . . .

Although Hawkins's enthusiastic restorations were quite well known among his contemporaries, it was not yet public knowledge. But all of that was about to change, and his restoration of the enormous ichthyosaur skeleton had repercussions that would reverberate all the way to the British House of Commons.

11

Mr. König Regrets

Dodging pedestrians and horse-drawn traffic, Mantell made his way through the Saturday afternoon crowds of summertime London in 1834, toward the familiar rendezvous on Charing Cross Road. Buckland never seemed to want to meet anywhere else, and always stayed at the Salopian Coffee House during his visits to London. But Mantell preferred other accommodations. After exchanging pleasantries the two men got down to the business of their meeting.

About a year before, Thomas Hawkins had offered his collection of marine reptiles for sale to the British Museum. His initial asking price of £4,000, which was an enormous sum back then—roughly equivalent to about $1 million—was summarily rejected, but Hawkins was not a man to be deterred. Even at twenty-three, he was wise to the ways of the world, and invoked Buckland's assistance to smooth the way ahead. Buckland, long since won over by the fulsome young fossilist, took up the matter with Charles König, the Keeper of Natural History, and the Reverend J. Forshall, Secretary of the British Museum. Hawkins showed his gratitude to Buckland in a fawning letter in which he expressed his willingness to cooperate in every way.

The following day Buckland, choosing his man carefully, wrote to Lord Farnborough, one of the trustees of the British Museum, expressing his view that this unique and truly outstanding collection must be secured for the nation. He then volunteered to evaluate the specimens on behalf of the museum, with Mantell's assistance. Buckland offered the opinion that the fossils, now stored in London, had an estimated value of between £1,000 and £1,500.

Within less than two weeks of receiving this letter, the British Museum received supporting letters from Clift, Conybeare, De la Beche, and Mantell, all at Buckland's behest. Buckland also wrote to the trustees, reiterating his willingness to act with Mantell in the valuation of the collection. He added that "The specimens offered for sale are beautifully and most accurately engraved in a folio volume, just published, by Mr. Hawkins . . . which may be considered as a catalogue of the collection." These words would later come back to haunt him.

Buckland and Mantell faced the age-old problem that dogs everyone involved in purchasing fossils: how to arrive at a fair price. Notwithstanding the exorbitant prices some fossils have fetched, like the $7.6 million recently paid for a *Tyrannosaurus rex* named Sue, one way to arrive at a fair value is to consider how much time and energy have gone into the final product. If I purchase an ichthyosaur skeleton from a Dorset collector for $60,000, it might sound like a lot of money. But consider the numbers. It would have taken many months of prospecting to find the specimen, perhaps a week to collect it, several months to prepare it, and probably several years before a comparable skeleton would be found again. Technicians are paid a salary of about $50,000 to do this work, so the purchase price is quite reasonable.

Buckland and Mantell were both experienced in buying fossils from local collectors, so they had a good idea of the prices fossils fetched. They agreed to work independently, keeping a tally of their estimated values for each specimen, and would compare notes afterward.

Hawkins's saurian collection was housed in a coach house on Ormond Mews, not far from the British Museum. There was also a second location on Adelaide Street, just off Charing Cross Road and close to the Salopian Coffee House. Neither location was an ideal storage place for fossils. Both buildings were dark and damp, and the specimens had been tightly packed in, without due consideration for accessibility. It was therefore difficult for Buckland and Mantell to see the specimens in their entirety, especially the large ones, like the splendid "Ichthyosaurus platyodon 25 feet long." But they managed somehow, and returned to the Salopian Coffee House, where they "sat up till nearly two o'clock and completed our valuation. . . . " Their appraisals were remarkably—suspiciously—close.

Their task completed, Mantell took his leave and repaired to his room at Osborne's Hotel. Buckland wrote two letters before retiring. The first was to the Trustees of the British Museum:

My Lords and Gentlemen, *London, 12 July 1834*

I have this day carefully looked over the collection of the Remains of Sauri offered for sale to the British Museum by Mr. Hawkins, and have had the valuable assistance of Mr. Mantell, in estimating every article separately. After carefully revising our notes, taken on separate lists, without communication with one another, we found our estimates of the whole to coincide within 5 £.; they are listed below:

All the specimens engraved in Mr. Hawkins's publication . . . we value at 1,025 £. The remainder of his collection of Sauri, not published in his work, but which we also strongly recommend to be purchased for the Museum, we value at 225 £. Total value, 1,250 £ . . . .

Buckland's second letter was to Hawkins:

Salopian Coffee House, 12 July, 1834.

I beg to return your catalogue, and with it enclose the amount of the valuation by Mr. Mantell and myself of your entire . . . collection of Remains of *Sauria*. I have sent a duplicate of this valuation to the Trustees of the British Museum. . . .

I have much satisfaction in telling you that when Mr. Mantell and myself compared our separate valuations . . . our estimates of the total valuation of each did not differ ten pounds . . . .

It is very difficult to see how two evaluators could arrive at such similar values—less than half a percentage point difference—if they were working independently. The first clue to how this may have been achieved is contained in a letter Hawkins wrote to Buckland just three days earlier:

Most anxious to effect the final disposition of my Collection . . . and conscious of the objections that a large sum of money for such kinds of purposes afford the economical. . . . I have sent to the Trustees, with my work, a list of all the *Sauri* . . . that I propose for the Museum, a copy of which I retain for you, *with the several prices as well as I remember that the articles have cost me, so that there may be no manner of mistake anywhere, and as little trouble in the estimation as possible.*

And, moreover, I take the liberty to express how much gratified I feel that *you and you alone determine the sum* that I am to receive . . . as I have not only every confidence in your judgement, but am sure that you will add all the more importance to the problem which you condescend to solve, *alone*. [My emphasis]

Hawkins clearly intended Buckland to take a leading role in his deliberations with Mantell, providing him with a list of the prices he expected to receive. Almost certainly unbeknown to Buckland, Hawkins had sent a somewhat similar letter to Mantell, ten days earlier. This unctuous letter, dated July 2, 1834, expressed the sender's sincere gratitude that the recipient had consented to join Buckland in evaluating his collection:

. . . two such excellent judges *cannot* fail to appreciate *justly* the value of any part of my Collection . . . and well knowing that objections might arise to the *amount* of the sum of money, I have set apart only indispensible [*sic*] specimens and estimating them with the spirit of a martyr make the gross amount of £2,300—a paltry sum for so grand an object surely. Now I do hope that you will do me the justice to believe that I rid myself of this part of my collection at the *very lowest* [multiple underlining] calculation for the sake of the Museum. . . . And I do entreat of you *(I speak with a freedom that I dare not use to Dr. B)* [double underlining] to advocate my estimate and enhance it as much as possible for why should this rich country rob one of one's own money. . . . I am perfectly justified being in *your* and the Doctor's hand . . . I leave the estimate to your care: as the doctor will not be dictated to by *me* or even influenced. When mine are fairly at the Museum, Oh how I should delight to see yours there also. Perhaps they may another day—and perchance *I* may be called upon to estimate them as you are mine. My dear Sir.

Ever most faithful and most obliged,

Thomas. Hawkins.
[Hawkins's underlining]

The manipulative Hawkins, having supplied both appraisers with the perceived value of his collection, led each one to believe that their's was the influential role in the final evaluation. Even more audacious, Hawkins attempted to influence Mantell into placing a high value on his collection by implying that he would one day return the favor, if the opportunity arose.

Charles König, as the Keeper of Natural History at the British Museum, had sole responsibility for the collections of mineralogy, paleontology, botany, and zoology. But he had no say in any of the major decisions affecting his department, such as the engagement and dismissal of staff, such matters being the prerogative of the trustees. Nor did he receive much assistance with his massive responsibilities. He worked hard, without ever managing to catch up on all the jobs he had to do. But he was uncomplaining. When he reorganized the mineral collection he preferred writing out all 12,000 specimen labels himself, rather than asking for assistance. The trustees never consulted him over the purchase of the Hawkins collection. However, the secretary did write to him to suggest he view the specimens, saying the trustees would likely ask his opinion. König followed Forshall's advice, though he appears to have given the collection only a fleeting inspection, largely because of its inaccessibility.

The overworked Keeper of Natural History had his first real look at the newly purchased Hawkins collection early the following year. Not counting the smallest items—isolated bones, teeth, coprolites (fossil faeces), and the like—there were over two dozen specimens. They ranged in size from a segment of jaw, well under one foot long, to the twenty-five-foot-long skeletons of *Ichthyosaur platyodon*. Many of them, including the latter, were wall-mounted in open wooden frames. They were to go on display in the public gallery, once glass-fronted display cases had been custom-built for them.

When the display case for the twenty-five-foot-long ichthyosaur was ready, König took time out from his busy schedule to make a careful inspection of the specimen, prior to its public debut. He probably stood and looked at the huge skeleton for some time, taking in all the details of its anatomy. The entire skeleton lay embedded in its gray liassic matrix, surrounded by a plaster in-filling, which helped secure it within the heavy wooden frame. The magnificent skeleton appeared to be complete, from the tip of its snout to the end of its tail. The four fins were complete, the ribs were complete—all the way down the length

The twenty-five-foot-long ichthyosaur skeleton that Hawkins restored, as it was when he it sold to the British Museum. It still looks the same today.

of the backbone—and the vertebral column was complete too. But surely the skeleton was not that perfect when it was discovered?

At this point König probably consulted a copy of Hawkins's book, comparing the illustrations of the specimen with the skeleton before him. The detailed lithograph depicted the right forefin only in outline, without any shading, showing it was missing. But in the skeleton the right forefin was as large as life, and looked every bit as real as its counterpart. Furthermore, the drawing depicted the vertebral column as ending abruptly, whereas the tail in the actual specimen was complete, right down to the minute vertebrae at its very tip. The legend at the bottom of the drawing read: "Redrawn from the Skeleton 18 feet long," whereas the mounted skeleton was closer to 25 feet in length. A new tail section must have been added, as well as a right forefin. But these additions looked just as real as the rest of the skeleton. König must have wondered how much more of the skeleton was artificial. The only way to be sure was to probe the entire skeleton with a sharp tool: Plaster, being much softer than bone, is easily detected by this simple test. He set to the task with a small knife, fearing the worst but hoping for the best.

Some time later König sat down at his desk and penned a letter to Buckland:

To Dr. Buckland

British Museum
Jan 20th 1835

My dear Sir,

I am sorry to be obliged to inform you that Mr. Hawkins's large specimen of Ichthyosaurus turns out to be made up to a degree of which you

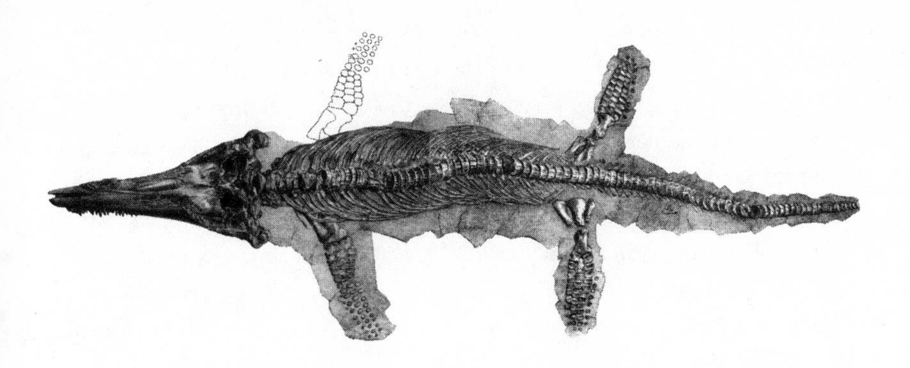

The twenty-five-foot-long ichthyosaur skeleton, as illustrated in Hawkins's book, where it is labeled as being only eighteen feet long.

> can have no conception without examining it *as it now appears* after having been touched all over with a knife to ascertain what is plaster and what bone: as to the soi-disant [so-called] lias, with all its fissures and rifts—it is entirely made up of plaster & lamp black. I shall not enter into details, & only write to beg you will let me know if you are likely soon to come to town, when you may see all with your own eyes. An expensive case has been made for this specimen, but the latter cannot be exhibited in its present state without manifest derogation from the character of the British Museum. Believe me, my dear sir
>
> *&c*
> *C K*

König wrote a similar letter to Mantell, but delayed informing the trustees for three weeks, presumably because he wanted to hear back from the two evaluators first. His formal report to the trustees, dated February 12, 1835, read:

> . . . Mr. König regrets he is necessitated to acquaint the Trustees that a discovery of rather vexatious nature has been made since the arrival of that [Hawkins] collection at the Museum. The two largest specimens, especially the principal Ichthyosaurus, which on account of its supposed perfection has been valued at 500 £., have turned out to be of much less value, on account of their being made up, and all over restored with plas-

ter of Paris, and altogether unfit to be exhibited to the public, without derogation from the character of the British Museum. Mr. K has thought it right to inform of this circumstance the two gentlemen upon whose recommendation and valuation the collection in question was purchased. Dr. Buckland, in answer, suggests that lines be drawn round the restored [parts], in order to distinguish them from the real osseous parts. Mr. Mantell, on the other hand, writes that if the specimens were his, he should let them remain as they are, and in his catalogue mention the parts that have been superadded or restored. . . . Under these circumstances nothing remains for Mr. König but to await the pleasure of the Trustees, as to the steps which are to be taken in this emergency.

König, who was not privy to Buckland's evaluation, was wrong about the value of the largest skeleton, having obtained the figure of £500 from the press, which had given a great deal of coverage to the purchase of the collection. The actual price was £200.

The trustees, communicating to König through the secretary, suggested the fabricated parts be painted a slightly different color. König concurred with this solution, thinking it better than removing the plaster restorations, which would make the specimen too short for its expensive glass case. In the meantime, Buckland visited the British Museum to examine the large skeleton for himself. He then wrote to the secretary. His letter warrants careful reading:

My dear Sir *Oxford, 12 March 1835*

In consequence of a letter from Mr. König, I called last week at the British Museum . . . I feel it due both to Mr. Hawkins and myself to request you . . . submit the following statement to the Trustees. . . .

That Mr. Mantell and myself made our estimates separately, and on comparing our lists found them to agree within 20 £. [previously reported as £5, and at £10] on the value of the whole collection:

That Mr. Hawkins never professed that there were no restorations . . . on the contrary, I was aware of what he had been doing: many of the specimens have for three or four years past been under my observation, and I have often remonstrated against a practice which I could not prevent. On more careful examination of the specimens, I find the amount of these restorations to be much less than I had supposed; and were I again to value the collection, I should fix a larger rather than a smaller price on it.

The principal restorations are in the largest specimen, which was valued only at 200 £ or 200 guineas; to obtain such a specimen in perfect state is all but impossible.

There has been, therefore, neither fraud nor collusion on the part of Mr. Hawkins, nor want of information on my part, as to the fact of reparation and restoration . . . and provided these restored parts be pointed out by a different colour . . . no one can possibly be deluded. . . .

As erroneous statements have appeared in the papers respecting this subject, you are welcome to make any use you think proper of this communication.

Mr. Hawkins would have done well had he indicated the amount of his restorations in his published plates; but this is a matter which affects the purchasers of his book, and not the Trustees of the Museum, who, being in possession of the specimen, can so readily remedy the existing evil by marking with a different colour the restorations . . .

Believe me to be, my dear Sir, always truly yours,

W. Buckland

Buckland would have been wise to have had his legal friend Lyell read the letter first, because he left himself wide open to litigation. In saying he had been aware of Hawkins's restorations for several years, and had remonstrated against him for it, he acknowledged his belief that the practice was wrong. He should therefore have been extra vigilant in the detection of restorations. His statement that his "more careful examination" reveals less reconstruction than he had supposed is a blatant admission that he did not carry out a careful inspection in the first place, though the inaccessibility of the specimens may have been a contributing factor. But his most damning admission of fault is the statement that Hawkins should have indicated the amount of restoration in his illustrations. Eight months previously, in his entreaty to the museum to purchase the collection, Buckland had stated, "The specimens offered for sale are beautifully and most accurately engraved in a folio volume, just published, by Mr. Hawkins. . . . " His argument that the only people deceived were the purchasers of Hawkins's book is false because the trustees had been deceived by it too.

The skeleton could certainly be rectified with differential coloration, but the point was that this should have been done in the first place. Hawkins clearly intended to give the impression that the large skeleton was complete. He did this by color-matching the restored parts to the original bone, and by fabricating artificial liassic matrix to make the specimen appear as if it were in its original state, as found in the rock.

Fortunately for Buckland the trustees of the British Museum were not litigious. Reassured by the letter from the eminent geologist, they were happy to smooth the matter over with the minimum amount of fuss. In addition to having the restored parts of the skeleton painted a slightly different color, a notice was placed in the gallery to the effect that these parts were artificial. The matter was quietly laid to rest.

Unfortunately for the museum, the Select Committee of the House of Commons was not content to let sleeping dogs lie. At the time of the purchase of the collection, the British Museum was undergoing a searching review by a special parliamentary committee. The museum had been subject to a long history of criticisms and misgivings regarding its operations and conditions, and the committee was charged with getting to the bottom of the matter. A number of witnesses were called to give testimony before the hearings, including many senior staff members and several outside specialists. Included in the latter group was the outspoken Scotsman, Robert Grant (1793–1874), surgeon, Lamarckian, sponge expert, and longtime holder of the Chair of Comparative Anatomy at London University. When asked how the zoological collection of the British Museum compared with its counterpart in Paris he told the committee that such a comparison was "almost ludicrous." In marked contrast to the French institution, the "Zoological department of the British Museum is miserable in funds, miserable in science, miserable in materials. . . . " The public galleries were a disgrace, and whereas the Paris Museum had dozens of magnificent stuffed animals on display, London's museum had only "some stuffed skins that once resembled [animals] . . . ," with "hideous rents" through which "we see straw and the beams of wood. . . . " He went on to say that the "lama's mouth is of painted plaster . . . the dentist has restored our skeleton of the elephant by thrusting not his own [teeth] into his jaws. . . . " He concluded that the painter, "with alternate daubs of red-lead and lamp-black, has with equal skill restored . . . our cassowary [bird]. . . . "

The committee had countless questions regarding the purchase of the Hawkins collection. König was closely interrogated but probably felt too intimidated by the courtroom formality to speak freely. Most of the questions related to the extent of the restorations, with several references to the likely intent of the fabrications, and their effect on the worth of the specimens. König was rather noncommittal in his responses. When asked whether he believed better evaluators could have been found, he said he thought not. König's views on the value of the collection were sought on several occasions. Although emphatic that the specimens were unique and of great intrinsic value to the British Museum, he was reluctant to give a monetary value. When asked whether he thought the collection was worth the £1,250 paid, he said it was a matter of opinion. His opinion was that it was rather too much.

The committee did not reach any conclusions, nor make any recommendations, but a transcript of the hearings was published by the House of Commons and entered into the public domain. Hawkins was outraged by the perceived imputations on his good name and character, and König was the butt of most of his anger. Many heated letters were exchanged between the two, with Hawkins supplying all the vitriol. Several years later Hawkins wrote a letter of protest to the trustees of the British Museum.

> [My collection was] an object of aversion to Mr. König as soon as offered in 1833, and of hatred when Prof. Buckland forced it upon him in 1834. No page of the history of Science contains a more disgraceful story than the Report of the House of Commons' Committee. . . . It exposes a furtive attempt of Mr. König to destroy . . . my moral character. . . . But I never reflect upon the condition of the famous large ichthyosaurus, and the manner in which it was handled, without a blush. . . . it [was] stabbed in every part, and it carries an inscription to this day which gives colour to the false charges. . . . [He railed against] an ostentatious announcement conspicuously placed on the fossil, "That all the coloured parts are restorations in plaster-of-Paris."—Can anything show more insatiable spite than this?

Hawkins pointed out that there were many other plaster casts in the galleries, but they were not similarly labeled. "I solemnly require . . . ," he continued, "in the name of Justice, and of my Country's behalf, the dismissal of all parties hitherto engaged about this noble Collection."

The large ichthyosaur skeleton at the center of the controversy can still be seen on public display in London's Natural History Museum. It is now referred to as the species *Temnodontosaurus platyodon,* but still has the same specimen number of 2003. Close inspection reveals that many of the bones have small pits, made by Mr. König's knife. And the extent of plaster restoration can be seen by the slight differences in color and surface texture. In addition to the right forefin and terminal segment of the tail, parts of many of the dorsal ribs (the ones forming the rib cage), all of the postsacral ribs (the short ribs that lie posterior to the pelvis), almost half of the left forefin, the tips of the two hindfins, and at least one of the vertebrae in the trunk region are also made of plaster. These plaster restorations were originally color-matched to the real bone and depicted in the illustration in Hawkins's book as if they were real. There can be little doubt that Hawkins intended to deceive the eye with his plaster restorations.

Mantell certainly felt deceived, as he wrote to his friend Silliman a year after his evaluation of the collection:

> The truth is there has been deception on his part. . . . Hawkins has . . . added many parts to his specimen—such as a paddle made of plaster, whereas one was wanting . . . had I been aware of this, although I should not have valued the collection at one dollar less than I did, yet I would have had these . . . either removed or pointed out to the Trustees. . . . the fact is he is non compos—it is the only excuse we[?] can make for his . . . impudence & folly. . . .

Hawkins was now living at Sharpham Park, former home of Henry Fielding, the author of *Tom Jones.* At this time the house and estate were rented to one Samuel Laver, who farmed the land. Laver, in his turn, sublet part of the house to Hawkins, a decision he regretted for the rest of his life.

The young fossilist, his head already filled with delusions of grandeur, took to his dignified surroundings like an earl to ermine. But he had barely settled in before he got into a bitter dispute with Laver. The disagreement, which would escalate over the next few years, appears to have begun with an incident over some strawberries.

One evening in early summer Samuel Laver was entertaining a young guest in his garden. Seeing some ripe strawberries, which the guest assumed belonged to Laver, the young man began helping himself. Moments later he was sent reeling by a resounding blow to the head. Turning around he saw Hawkins, in a white rage, claiming the young man had stolen his strawberries. A heated exchange ensued between Hawkins and Laver, sparking a feud that would never be resolved.

At about the time of all this unpleasantness in Somerset, Edward Charlesworth, the editor of the popular *Magazine of Natural History,* was sitting down to a small dinner party at the London home of a close friend. During the dinner conversation the topic turned to fossils, and the name of Hawkins was raised by some of the other guests. Charlesworth's opinion was sought on the affair with the British Museum, which had been in all the papers. Charlesworth, a Fellow of the Geological Society, had worked as a museum curator, including a short stint at the British Museum in 1836, and had taken a particular interest in the story, ever since Hawkins took him to see his collection prior to its sale. Charlesworth had read the Parliamentary Report in detail, and was of the opinion that the museum had been deceived. When he voiced this opinion he discovered that the guests who asked the question were personal friends of Hawkins. And they did not like what they heard.

Shortly after the dinner party Charlesworth received a letter from one of the guests, inviting him to retract or justify his opinions of Hawkins. When Charlesworth wrote back declining to retract he had no idea that legal proceedings were being contemplated. But this became apparent when he received a second letter, dated December 31, 1838. This letter was from Robert Young, another friend of Hawkins who was at the dinner party. He told Charlesworth that Hawkins was prepared to suspend proceedings against him for a week, but that he required terms for a settlement of the matter. To this end Young invited Charlesworth to a breakfast meeting the next day.

Charlesworth kept the appointment. He remained convinced that Hawkins had no legitimate case against him, however, and steadfastly refused to retract his remarks. Hoping to persuade him otherwise, Young sent him copies of two letters he had received from Hawkins. In the first letter, dated December 29, 1838, Hawkins announced his intention to:

> subpoena every witness that ever had anything to do with my transactions with the British Museum. . . . [He said it was his] duty to crush for

> ever a lie. . . . This is the only object I have—to crush this lie for ever—it can be effected only by a public retraction, that is all I demand, and that I will, so help me God, obtain.

He gave Charlesworth a week's grace to write "an ample apology," before instructing his solicitor to act. In the second letter, dated six days later, Hawkins granted Charlesworth, "the further grace of three days," to write "that self-damning confession which I will move heaven and earth to have. . . ."

Far from intimidating Charlesworth into writing a retraction, these letters only confirmed his suspicion that Hawkins had no case. Had it been otherwise, Hawkins would have instructed his solicitor to proceed when Charlesworth refused to comply. Charlesworth, who had an argumentative streak himself, instructed his lawyer to write to Hawkins. The letter informed the complainant that there would be neither an apology nor a retraction of Charlesworth's comments, and that they were prepared to take all necessary steps for meeting whatever legal proceedings they thought appropriate.

At length Hawkins sent a rambling declaration to Charlesworth's solicitor, laying damages at £1,000 and accusing Charlesworth of "contriving and wickedly intending to injure him in his said good name, fame and credit, and to bring him into public scandal, infamy, and disgrace, and to vex, harass, and oppress the plaintiff. . . . "

Hawkins, not a man to miss an opportunity, involved Buckland in the dispute. When Lyell called on his old professor, Buckland told him that the editor of the *Magazine of Natural History* was on "the brink of ruin." After apprising Lyell of the situation, Buckland asked him to try to persuade Charlesworth to comply with Hawkins's demands. Lyell willingly called on Charlesworth in an attempt to do this, but was unsuccessful. Charlesworth appreciated Lyell's good intentions, but he was most aggrieved at Buckland for telling a third party, especially one who was a contributor to his magazine, that he was on the brink of ruin, merely because of a pending action by Hawkins.

Some time passed, then Lyell wrote to Charlesworth, telling him that Buckland was "most willing to be a mediator," and that he expected "you to give a verbal apology of some sort before the parties in whose company you used the expressions complained of. . . . " Charlesworth had no sooner declined Buckland's offer of mediation when he received a letter from the Reverend J. Forshall, secretary of the British Museum:

Dear Charlesworth, *18 Feb. 1839*

Dr. Buckland has called upon me in reference to some dispute between you and Mr. Hawkins. Dr. Buckland seems to think that you are ignorant of the facts of the case, and that I may possibly be able to explain them to you.

Have you any objection to come and see me?

Yours very truly,

J. Forshall.

Charlesworth had a high regard for Forshall and readily agreed to a meeting, but nothing was changed by their discussion.

Five or six weeks passed uneventfully, leading Charlesworth to conclude that Hawkins had given up. But then he received another visit from Lyell. Lyell told him that at the previous night's meeting of the Geological Society, Buckland had let it be known that he intended to join Hawkins in the action, urging Hawkins not to let the matter drop.

Charlesworth suspected that Buckland had no such intention, and this was confirmed a few days later when a certain document fell into his hands. The document revealed "that the affair is suspended for some time . . . the delay is tantamount to bringing no action at all. . . . It moreover appears that Buckland wrote to Hawkins a letter, advising him . . . not to proceed . . . ."

Charlesworth was so incensed by the entire business, especially by the actions of a man of Buckland's stature, that he published an account of the whole affair in an appendix of the *Magazine of Natural History.* Citing relevant extracts from the parliamentary report, together with certain letters pertinent to the case, he documented the reasons why he felt the museum had been deceived. By documenting the many inconsistencies in Buckland's actions and statements, he also showed how incompetent he had been in evaluating the Hawkins collection. He concluded with a withering blast at Buckland:

> I *still* think the Trustees of the British Museum were not fairly treated in the purchase of Mr. Hawkins's fossils, and if Dr. Buckland deems it advisable to prosecute me for thinking so, he can make a cat's paw of his friend. . . . He cannot get into the witness-box with clean hands, after

> one day volunteering to mediate for me, and the next to change places with Mr. Hawkins and become my prosecutor. Nor do I believe . . . a jury willing to attach weight to any statement he might depose . . . after the duplicity which I have shown him to be capable of . . .

Charlesworth heard nothing more on the matter, either from Buckland or Hawkins.

12

The Professor and the Naturalist

While Hawkins was nursing his wounds, Richard Owen was rising quickly in the rarefied world of academia. Luck, and the approbation of influential men like Buckland, probably played some small measure in Owen's rapid ascent. But most of his success was attributable to his own brilliance and sheer hard work.

Owen, like Mantell, began as a physician, completing his studies at London's Royal College of Surgeons under the same John Abernethy (1764–1831) who was both president of the college and of the college's Hunterian Museum. Abernethy, blunt to the point of rudeness and not easily impressed, was much impressed by the anatomical competence of the young man from Lancaster. He made Owen his prosector—the one who dissects the body before the anatomy class—and when he finished his studies and set up a small medical practice, Abernethy offered him a position at the museum. The annual salary of £120 was modest, even for those times, but Owen had the small income from his medical practice and happily accepted.

Named after John Hunter (1728–1793), the celebrated surgeon, the Hunterian Museum housed Hunter's huge natural history collection, which included animals preserved in spirits, skeletons, and fossils. Most of the preserved specimens had been collected by Joseph Banks during Captain Cook's voyage to the South Pacific (1768–1771). The collection had been under the care of Sir Everard Home, but he had done no curating in over twenty-five years, much to Abernethy's chagrin. Owen was charged with putting matters right, and the industrious young man from Lancashire set to work organizing and cataloging the specimens with a will. He also dissected a wide variety of

Richard Owen, as a young man.

animals, both those with backbones and those without, thereby broadening his knowledge of comparative anatomy.

Owen was engaged as an assistant to William Clift, the conservator of the museum, a position that would today be called chief curator or director. Clift's son, another William, was also an assistant and was due to take over his father's position when he retired. Clift senior, a highly respected anatomist and Fellow of the Royal Society, seems to have taken an instant liking to Owen, and Owen to him.

Clift had a daughter, Caroline. She and Owen fell in love and began an eight-year courtship, imposed by Mrs. Clift's insistence that Owen should have sufficient means to support her. William Clift wholeheartedly approved of the match, and bent to the task of furthering the career of the promising young anatomist. Owen obtained access to animals that died at the London Zoological Gardens, doubtless through Clift's influence, and began to publish

The public gallery of the Hunterian Museum, from a drawing of 1845.

a series of anatomical studies, beginning with one on the orangutan. He also began to publish a series of catalogs of the Hunterian collection. Owen spent long days in the laboratory and late nights at his desk.

Owen did not confine his researches to vertebrates during the early years of his career. One of his major works was a detailed study of the anatomy of nautilus, a living relative of the ammonites. He sent a draft of the paper to Buckland, who was greatly impressed by the meticulous quality of his work. It was a shrewd move on Owen's part to have the approbation of a man of such influence. Clift must have been well pleased with the way his future son-in-law's career was shaping, and everything seemed right and well with the world. But all of that took a turn for the worst when his son was involved in a tragic accident while returning home one night by cab. The cab driver took a turn too fast, overturning the vehicle and throwing his passenger onto the road. Young Clift struck his head with such violence that he was knocked unconscious. The insensible young man was rushed to St. Bartholomew's Hospital, where, as fate would have it, he was attended to by Richard Owen. Owen's examination revealed a fractured skull—a mortal injury. Clift senior was on holiday at the time and was difficult to contact. When word finally reached him he rushed home, just in time to see his son die. Owen was now the only museum assistant, and his salary was increased accordingly. It is an ill wind that blows no good.

Owen's teaching career at the Royal College of Surgeons began in 1836 with his appointment to the prestigious post of Hunterian Professor of Anatomy. He was still only thirty-three, and was enormously gratified by the singular honor. His teaching duties did not commence for ten months, but he used the time wisely in the interim to prepare his entire lecture series for the year.

Owen already had three years' teaching experience at the time of his inaugural lecture, but it was still an anxious man who faced the capacity audience on that May afternoon in 1837. He had reason to be nervous. The Oxford gentlemen Buckland taught attended classes largely from interest and were quite undemanding. But the London students Owen had to face were required to attend classes, and if they did not get what they expected their disapproval could be overtly dangerous. Just five years earlier, the students, dissatisfied with the lecture, broke out into a full-scale riot.

The previous night he had read the one-hour lecture to Caroline—Carey, as he called his wife of one year. It ran for ninety minutes. That was an hour shorter than on his first reading several months before, but it was still far too long. He worked on into the early hours, cutting and moving some of the material to the second lecture.

"So busy all the morning," Caroline wrote in her diary on the fateful day, "had hardly time to be nervous. . . . R. robed in the drawing-room and took some egg and wine before going into the theatre. He then went in and left me." The couple lived in a garret apartment provided by the college. This over-

The lecture theater of the Royal College of Surgeons, as it would have looked at the time of Owen's inaugural lecture. The lecturer shown in this 1844 illustration is not Owen.

looked the ground-floor lecture theater, so if rioting did break out among the students, she would know about it as soon as her husband.

Riots notwithstanding, lectures at the Royal College of Surgeons were every bit as formal as they were at Oxford or Cambridge. Students and staff took their places first, and as the clock struck four the mace-bearer appeared, followed by the president, his council, and their honored guests. The 400 people crammed into the theater to hear his inaugural address included members of Parliament and some of the most senior men of science and medicine in the land. And into that hushed and expectant place came the Lancastrian lamb for slaughter.

After some preliminary remarks on the gravity of his new position and the importance of the teaching duties with which he had been charged, he gave a brief overview of the series of twenty-four lectures he would deliver. According to stipulation, these had to be illustrative of Hunter's collection. Having respectfully acknowledged the celebrated surgeon's contributions to comparative anatomy, Owen launched into a history of natural science. Caroline wrote about the event in her journal:

> After 5 o'clock a great noise of clapping made me jump. . . . All went off as well as even I could wish. . . . R. was more collected than he ever supposed, and gave this awful first lecture almost to his own satisfaction! We sat down a large party to dinner.
>
> Mr. Langshaw and R. afterwards played two of Corelli's sonatas. [Owen, an accomplished musician, played the viola.]

If his listeners were pleased with his first lecture, they must have been enthralled by the second one, two days later. Owen took his audience back to the time of the ancients. He applauded the Greeks for the scope and depths of their zoological enquiries, from the classification of animals to internal parasites that caused sickness. Aristotle, Owen told his attentive audience, correlated an animal's lifestyle with its anatomy. He observed, for example, that carnivores had sharp and pointed teeth, for cutting, while those of herbivores were flat, for grinding. Owen marveled at the way that Aristotle, without even a hand lens to assist him, had observed the embryonic development of the chicken. But Aristotle's most significant contribution was his recognition of homology.

The principle of homology recognizes the equivalence of a certain part of the body of one animal with that of another animal. The examples Owen gave included a lion's claw with a horse's hoof, and the wing of a bird with the forelimb of a mammal. Although a bird's wing looks very different from our own arm, they are built on a common plan, from similar bones with similar connections and developmental origins. The major differences are in the shapes and relative proportions of the individual bones.

Owen criticized some contemporary anatomists for essentially repeating the Greeks' work on homology, "Believing they were working out an intirely [*sic*] new principle. . . . "With all the audacity in the world they had "*dignified* their speculations with the Epithet Transcendental! . . . implying theirs' to be the only *philosophical* mode of considering the Subject."

The point Owen wanted to drive home was that the "new" theories conceived on the Continent, which were becoming so fashionable among the avant-garde in Edinburgh and London, were nothing new at all. They were merely extensions of what the Greeks had conceived 2,000 years before. He was especially critical of the sweeping theories of people like Geoffroy (Etienne Geoffroy Saint-Hilaire, 1772–1844). According to Geoffroy's "philosophical anatomy" or "transcendental morphology," the development of all animals was restricted to a common body plan. It followed from this that it should be possible to find homologies for every single bone in a vertebrate's skeleton. Cuvier, Geoffroy's colleague at the natural history museum in Paris, disagreed. His own research on the vertebrate skull had shown, for example, that although most of the bones of a fish were homologous with those of a mammal, there were no homologies for the bones of a fish's operculum (gill flap). This was no surprise to Cuvier because he considered structures were determined by their function rather than by any constraint on their development. As fish were the only vertebrates that used specialized gill slits to breathe, it was inevitable that their opercular bones should be unique too.

According to Geoffroy's theory, such homologies *had* to exist. Geoffroy was therefore determined to find some matches for the opercular bones in the mammalian skull. He searched in vain. Then he realized that Cuvier had not matched up the three small internal bones of a mammal's ear with any of the head bones of a fish—these are the small bones that link the eardrum with the inner ear. Here, at last, Geoffroy proclaimed, were the missing homologies for the opercular bones. But he was wrong, as Owen told his audience, and such grave errors were "the result of an abuse of a sound and fruitful Principle [homology]. . . . " Simply put, Geoffroy was giving homology a bad name! Some transcendentalists went so far as to attempt to homologize structures between vertebrate and invertebrate animals. The claw of a lobster, for example, might be homologized with the human hand, which was patently absurd.

Having castigated some of his Continental contemporaries, Owen returned to his theme of the ancients. He contrasted the achievements of the Greeks with the wasted opportunities of the Romans. The sadistic circuses they held in their coliseums resulted in the slaughter of countless exotic animals, but none appear to have been dissected afterward. Owen recounted how a great Roman victory over the Carthaginians was celebrated by the killing of 142 captured elephants with bows and arrows. On another occasion a large number of ostriches were released into the arena. As the terrified birds ran about, their heads were shot off with arrows, specially fitted with crescent-shaped blades for the purpose.

Owen's lectures were an unqualified success. They won the respect of his students, the praise and admiration of his peers, and the attention of the most influential people in the land. Indeed, the lectures probably did more for the flourishing career of the brilliant young anatomist than even his prodigious output of scholarly publications. Buckland came all the way from Oxford to hear Owen, and while he sat and listened to the ascending star, their respective wives took tea together and chatted. Lyell attended too, and was most complimentary about Owen's clear and distinctive voice, which could be plainly heard by everyone in the room. The *London Medical and Surgical Journal,* traditionally hostile toward the Royal College of Surgeons, wrote that the lectures were in striking contrast to the "morbid ossific compositions which immediately preceded them. . . . "

Owen was the man of the moment. The Royal Institution elected him as their Fullerian Professor of Comparative Anatomy and Physiology. Although flattered by the honor, Owen was obliged to decline the position by his own college—they did not want him taking on any more commitments until the catalog of Hunter's collection had been completed.

Owen gave three lectures each week, on Tuesdays, Thursdays, and Saturdays. It was a grueling pace, and their preparation was an enormous drain on his time and energy. "R. up till two this morning writing," Caroline noted after his second lecture. By the third week he was feeling sufficiently confident to lecture without notes. When he apologized for the fact, he received two rounds of applause. However, by the midpoint in the series he was showing signs of strain. "R. very queer on coming back from lecture," Caroline wrote, "if he is not better by next lecture I shall try and get it postponed." But her husband carried on without interruption.

Owen would continue with the series for the next two decades, without ever repeating a single lecture. Anyone familiar with teaching knows what a remarkable achievement this is. The lectures ended only when he relinquished his teaching position to become the Director of the British Museum.

We can learn a great deal about Owen's views of the natural world from his lectures. In his fourth lecture, for example, he discussed the works of embryologists, and how "higher" animals, like mammals, appeared to go through stages during their development like those of "lower" forms, such as fish. But he warned his audience that the various structures seen during the development of one animal should not be equated with corresponding structures seen among adults of other species. He did not give specific examples, but one of the obvious ones he might have chosen was the gill slits. Adult fish have gill slits through which they breathe, and these are formed during their embryonic development. Neither reptiles, birds, nor mammals have gill slits as adults. However, there is a transitory phase during their early embryonic development when gill slits appear. Owen's point was that the embryonic gill slits of a monkey or a lion should not be equated with the adult gill slits of a cod or a herring.

Owen's treatment of embryos may have seemed esoteric and irrelevant to his students. However, he warned them that the transformation of embryos during their development was: "closely allied to that still more objectionable one, the transmutation of Species. Both propositions are crushed in an instant when disrobed . . . and examined by the light of severe logic."

Owen was opposed to the idea of the transmutation of species because it went directly against his Christian beliefs in the creative power of God, beliefs he held throughout his life. In later years (1864), he would write a short book on God and creation, based on a lecture he gave to the Young Men's Christian Association.

Owen returned to the question of the transmutation of species in his fifth lecture, citing the by now familiar example of mummified animals in Egyptian tombs as evidence for the permanence of species. He concluded by discounting that the sequential appearance of different kinds of organisms throughout geological time was a consequence of transmutation (evolution): "There is not a single fact which demonstrates . . . that the changes of form which we witness in the succession of the organized inhabitants of the earth were produced by progressive development, or transmutation of Specific Forms."

Following Cuvier's reasoning, Owen attributed the apparent succession of life seen in the fossil record to repeated waves of creation and extinction. These, in turn, were attributable to the successive revolutions of physical change that had swept the globe.

Owen made the point that the new species that appeared in the geological sequence were "not necessarily superior . . . to the extinct species which they have replaced." Here was his major objection to the idea that the fossil record was progressive. As an example of this he cited the case of modern cephalopods, such as the squid, which he considered were "a much lower Type of Molluscous organization" than the ammonites of the remote past. He accordingly rejected the idea that the fossil record showed a trend from "lower" to "higher" animals and plants, a position that Lyell had taken in *Principles of Geology*.

Owen was an exceptionally prolific writer. At the time of his inaugural lecture he already had about sixty publications in print—more than most modern anatomists achieve in a lifetime. He would continue to publish until his eighty-fifth year, producing over 600 papers, a monumental achievement. His first few papers were related to medicine, but these soon gave way to a string of anatomical studies on a diverse array of animals, ranging from kiwis and clams to tigers and tapeworms. The order in which these publications appeared was largely determined by mortalities at the Zoological Gardens. The Owens usually visited the zoo each Sunday after church, always paying particular attention to the health of the animals. There are several pertinent entries in Caroline's diary, including: "the poor little chimpanzee is lying so ill" and "The poor lion lying in straw and almost dead."

Rising well before dawn one summer Sunday, they set off to the zoo to witness the arrival of four new giraffes. The procession of "graceful, bounding playful giraffes, attended by . . . four Africans in native costume" had walked eight miles from the docks at Blackwall on the river Thames. Six months later Owen dissected one of the giraffes after its untimely death.

The London zoo was not his only source of supply of carcasses. When a rhinoceros died at Wombwell's Menagerie the carcass was delivered to Owen's home. His wife recorded that "The defunct rhinoceros arrived while R. was out. I told the men to take it right to the end of the long passage, where it now lies. As yet I feel indifferent, but when the pie is opened—."

Cataloging the Hunterian Museum's 20,000 specimens was an enormous undertaking, especially on top of Owen's teaching responsibilities. The only way he could maintain his prolific publication output was by working into the early hours, and by phasing out his medical practice. In spite of his grueling work schedule, he still managed to squeeze in some recreational time with his wife, their son Richard, and his wife's family. They both enjoyed the theater and music, and often went to concerts and to plays, sometimes just the two of them, and sometimes with Caroline's parents. There were frequent musical evenings at home too, with Caroline at the piano and her husband, who also had a fine singing voice, playing the viola. He read for relaxation as well as for interest, and was especially fond of Dickens, an author he greatly admired. He read *Sketches by Boz* as soon as they appeared in print, and devoured all of Dickens's novels. He eventually met his literary idol at a theater gala attended by the young Queen Victoria, and the two men became good friends. Dickens even mentioned Owen in *Our Mutual Friend,* where Mrs. Podsnap is described as being a "fine woman for Professor Owen," on account of her interesting anatomical features.

Owen's extensive knowledge of the anatomy of modern animals gave him a considerable advantage when it came to interpreting extinct ones. His first paleontological work, which appeared in 1837, was on a fossil mammal brought back from Argentina by Charles Darwin (1809–1882), during his voyage aboard the *Beagle*.

Owen's introduction to Darwin, and his fossils, began with a letter from Lyell.

> *16 Hart Street, Bloomsbury:*
> *October 26, 1836*
>
> My dear Sir,—Mrs. Lyell and I expect a few friends here on Saturday next, 29th, to an early tea party at eight o'clock, and it will give us great pleasure if you can join it.
>
> Among others you will meet Mr. Charles Darwin . . . just returned from South America, where he has laboured for zoologists as well as for hammer-bearers. I have also asked your friend Broderip.
>
> Yours faithfully,
>
> *Charles Lyell.*

At this time the *Beagle,* Darwin's home during the previous five years of circumnavigating the globe, lay berthed at Woolwich dockyard. Her crew would soon be paid off and go their separate ways, to new ships and new duties in the services of the king. But Charles Darwin, twenty-seven years old, unmarried, unemployed, but not without means, was readjusting to life back on land. He had been dividing his time between the family home in Shrewsbury, his old alma mater in Cambridge, and London. Darwin detested the city, "this London is a vile smoky place," but it was a means to an end—he had to find specialists to work on the large collections he had brought back from his voyage, and London was his best prospect.

The tea party Owen attended at Lyell's London home on that late autumn evening would have been a most agreeable occasion. Their urbane host, congenial and poised, accompanied by his lovely wife, would have put everyone at their ease. Darwin, with a thousand tales to tell, likely entertained and enthralled the other guests. Well aware of Owen's distinguished reputation, he probably deferred to him on matters zoological. Owen, five years his senior, might have revealed a hint of the arrogance that would become so much a part of his character in later years. Owen's friend, the magistrate William Broderip, was an accomplished malacologist, with a large collection of shells in his Lincoln's Inn chambers. Darwin's descriptions of the exotic shells he had picked up on tropical beaches around the world would have thrilled Broderip. He happily agreed to look over the South American shell collection, and Owen seemed "anxious to dissect some of the animals in spirits."

The evening was a tremendous success, especially for Darwin. Not only had he managed to farm out some of his specimens for study, but he had made a good friend in Lyell. "Amongst the great scientific men," Darwin wrote of him a few weeks later, "no one has been nearly so friendly and kind as Lyell."

At this time Darwin was especially interested in geology, and wrote several papers on the subject. He read his first paper at the Geological Society on January 4, 1837. As a young novice among so many eminent men he felt nervous and overawed, but his insecurity was groundless. As he wrote to a friend later that year: "I have read some short papers to the Geological Society, and they were favourably received by the great guns . . . this gives me much confidence, and I hope not a very great deal of vanity, though I confess I feel too often like a peacock admiring his tail."

Lyell was especially interested in Darwin's views on geological processes, and was much impressed by his ideas on how coral islands were formed. "I am very full of Darwin's new theory of Coral Islands, and have urged Whewell to make him read it at our next meeting." Darwin, for his part, was every bit as full of Lyell: "I have a capital friend in Lyell, and see a great deal of him, which is very advantageous to me in discussing much South American geology."

Darwin moved to London at the end of 1836, and was busily writing up his account of the voyage of the *Beagle*. He was still trying to find specialists to study parts of his collection, a task that had proved to be as difficult as some of his friends had warned him. There was no shortage of collectors, but precious few of them were competent naturalists. He needed men who could identify the specimens that were known to science, describe those that were new, and to do this expeditiously. Such men did exist, but they were too busy with other things. John Gould, the painter and naturalist, took the birds off his hands, and would later describe the remarkable finches that came to bear Darwin's name. Buckland agreed to look at some of the fossil bones, but most of them were sent to the Hunterian Museum for Owen's attention.

Darwin paid a visit to Owen during the first week of December 1836, and delivered a consignment of fossils. The museum had been closed for the previous two years for extensive renovations, and was still a clutter of bare walls and plaster, with dust and builders everywhere. Darwin and Owen only added to the chaos as they pried open shipping crates, strewing packing material and fossils everywhere.

Darwin paid many more visits to Owen, both socially and professionally: "Mr. Darwin here this afternoon," Caroline Owen wrote in her diary the following spring. "After tea muscular fibre and microscope in the drawing-room." On another visit, during the summer of 1837, Owen and Darwin spent a morning together examining and discussing the South American fossils. Owen had begun to work on them as soon as they were delivered the previous December, and had already got names for the new species. Writing to Lyell after the visit, Darwin told him that "We made out the remains of 11 or 12 great animals. . . . "

Most of the fossils were new to science. One of them, a large skull of a rhinoceros-sized beast, was especially intriguing because it appeared to be a cross between a rodent and an elephant! Owen decided to call the strange animal *Toxodon,* meaning "bow tooth," for the distinctive shape of its teeth. The paper in which he described it was his first paleontological publication. A second new animal was about the size of a camel and had a long neck. Owen thought it was related to the llama, and named it *Macrauchenia,* meaning "long neck." *Macrauchenia* was unusual for having its nostrils on top of its head like an elephant, suggesting that it too may have had some sort of trunk. Although *Toxodon* and *Macrauchenia* appeared to be unrelated at the time, we now recognize that they belong to the same South American group of mammals called notoungulates. Perhaps the most remarkable of the discoveries was a hippopotamus-sized animal that Owen named *Glyptodon,* meaning "sculptured tooth." What made it so unusual was that its body was encased in a bony shell, like that of a tortoise. But it was a mammal, related to the modern armadillo.

Owen identified some of the fossils as giant ground sloths, most of which were already known to science. But he did discover an entirely new kind, which he named *Scelidotherium,* meaning "limb beast." The giant sloths were obviously similar to their diminutive cousins still living in South America, but some were as large as elephants.

Darwin knew very little about fossils himself, and was therefore completely dependent on Owen for their interpretation. He probably hung on every word Owen had to say about the bizarre creatures. Among the many things that intrigued the young naturalist was the cause of their extinction. Owen

The South American fossil *Glyptodon,* a mammalian relative of the armadillo.

attributed this to changes in the environment—some new revolution that had swept the world—but Darwin was not convinced. In a letter to Lyell, written a few days after the morning spent with Owen and the fossils, Darwin commented: "What an extraordinary mystery it is, the cause of death of these numerous animals, so recently, & with so little physical change [in the environment]." Two years later he would write: "Since their loss, no very great physical changes can have taken place in the nature of the country. What then has exterminated so many living creatures? In the Pampas, the great sepulchre of such remains, there are no signs of violence, but on the contrary, of the most quiet sensible changes."

Here we see Darwin questioning Cuvier's widely held catastrophic explanation for extinction, as advocated by Owen and Buckland. But what interested Darwin most about these strange South American fossils was their obvious affinities with animals still living there today. In his autobiography, written almost two decades after the *Origin of Species* (1859), Darwin cited these fossils as one of the key pieces of evidence that made him realize species were not immutable. It is probably no coincidence that Darwin started to write his first notebook on the question of the origin of new species at about the time he discussed the South American fossils with Owen. Imagine Owen's reaction had he known of the transmutational thoughts running through Darwin's mind as they picked over fossils together.

As Darwin wrestled with the species problem, he discussed his thoughts with Lyell and a small circle of trusted confidants. Darwin did not choose Lyell because Lyell was sympathetic to the idea that species were mutable—quite the reverse was true—but because he knew he could be trusted not to divulge anything of their discussions. There was good reason for such secrecy. In those

Owen's restoration of the giant ground sloth *Mylodon*.

times, the suspicion of being a transmutationist was every bit as damning as the suspicion of being a communist during the McCarthy era in the United States. Neither Owen nor any of the others in the geological circle, save a select few, would learn of Darwin's heresy until the year before publication of the *Origin of Species.* Darwin remained a clandestine asp among the angels.

Owen's work on the South American fossils was acknowledged by the Geological Society's most prestigious honor, the Wollaston Gold Medal, awarded to him in 1838. During the presentation of the medal, the president of the society made a commendatory remark about Darwin, saying his voyage aboard the *Beagle* was one of the most important events for geology in many years. What the president did not say, because he did not know, was that the voyage had set Darwin on a course to revolutionize the way life on the Earth would be perceived. And the greatest irony in all this was that Owen, the arch-conservative and staunch proponent of the immutability of species, contributed so much to Darwin's transmutational cause with his work on the South American fossils.

Darwin's biological knowledge was primarily acquired by observing living things in nature. Owen, in contrast, learned about animals by studying their mortal remains in the laboratory. Although he was not a field man, Owen did take the occasional excursion, usually to attend meetings or to study other museum collections. During a visit to York he took a train journey to Derby, and gave a vivid description of hurtling through pitch-black tunnels, with sparks flying and the incessant shriek of the steam engine's whistle. His companion, Lord Enniskillen, who was even taller than Owen, insisted on riding outside, causing Owen great concern for his safety. Each time the clanking engine lurched into daylight Owen stretched out of the window "to catch a glimpse of his head, if still in the right place . . . "

One of Owen's most singular visits was to the studio of J. M. W. Turner, accompanied by his friend Broderip. The artist, whose ethereal images of sail and steam distilled the essence of the changing times, lived off London's fashionable Harley Street. After a long wait on the doorstep of Turner's somewhat dilapidated house, they were admitted by a wary old woman who escorted them into a room with shuttered windows. When the door closed behind them they were surprised to find themselves in complete darkness. At length the old woman returned, and escorted them to the studio at the top of the house. Turner was standing before several canvases, which he was painting at the same time, taking his pigments from a rotating round tabletop. He showed them all his paintings, explaining that the reason for the blackout was so their eyes could appreciate them properly.

After attending the 1839 British Association meeting in Bristol, Owen traveled to the West Country, first stopping to visit Thomas Hawkins, near Glastonbury. Hawkins served him peacocks' eggs for breakfast, and introduced him to all his neighbors. By some mistake the neighbors had confused him with Professor Robert Owen, the social reformer. At length the error was corrected, and Owen "established at length my claims to be regarded as one of the same species. . . . "

His next stop was Lyme Regis, where he met Buckland and Conybeare. "They made me prisoner, and drove me off to Axminster, of which Conybeare is rector. Next day we had a geological excursion with Mary Anning, and had like to have been swamped by the tide. We were cut off from rounding a point, and had to scramble over the cliffs."

Owen returned to London well before the winter of 1839 set in. This is a great pity because if he had remained in Lyme during late December, he would have witnessed one of the most remarkable geological catastrophes of the age.

13

Landslides, Glaciers, and Riots

Most of the inhabitants of Lyme had taken to their beds well before midnight on Christmas night, 1839. But some lamps and candles burned on, shedding a little light onto the deserted streets. Gooses had been eaten, presents opened, and toasts drunk for another year—at least in the wealthy homes. Buckland's light was probably one of those still burning—the Bucklands were staying in town for the holiday season. He planned to do a little fossiling and see some local collectors. They would also pay a visit to Conybeare, now living in Axminster, just five miles from Lyme. Buckland always enjoyed his winter breaks away from Oxford. Sometimes the weather could be cooperative, gracing the Dorset coast with short days of golden winter sunshine, but the nights were usually cold. That year's autumn had been exceptionally wet, even by English standards, and everyone had grown weary of gray skies and galoshes. If the weather only stayed dry he would be more than content. We can picture him putting another log on the fire, refilling his glass, and sitting back in tranquil contemplation of the days to come.

Three miles along the coast, west of the Cobb, west of Monmouth Beach, and west of Pinhay Bay, were a pair of cottages. They sat perched atop the Undercliff—an elevated strip of land some half-mile wide that abutted the cliff. The cottagers had a commanding view of the sea, and could often see the hazy ghost of the French coast on the other side of the Channel. But there was nothing to be seen on that black night. The cottages themselves were in darkness too, their respective occupants having ended their modest celebrations some time before, and taken to their beds.

Two men picked their careful way across the pebbles, mindful of where they walked and the need for silence. They kept vigilant watch about them, alert to any movements or sounds, both along the beach and out to sea. Smuggling had been a local industry along this stretch of coast for as long as anyone could remember, and their job as excise men was to discourage the practice. They received neither support nor succor from the locals, who considered smuggling almost their birthright. Mary Anning herself sometimes found the occasional keg of contraband along the shore, and when she did she always covered it up to hide it from the excisemen. She would then tell some deserving family where to find the hidden cache.

At about midnight the inhabitants of the two cottages were startled into wakefulness by a thunderous roar that could be felt as much as heard, one that shook them in their beds with great violence. It was as if their homes were being torn apart by some unimaginable force. When they opened their eyes and saw "the floors of their houses rising upward towards the ceiling . . . ," imagination became a terrifying reality.

While the terrified cottagers were scrambling clear of their wrecked homes, the two excise men witnessed a scene that confounded their comprehension. Standing in complete disbelief, they watched "a huge reef of rocks gradually rising out of the sea at a short distance from the shore." The reef would continue to rise the following day, reaching heights up to forty feet and stretching half a mile along the coast. Fearful for their lives, and not knowing what to expect next, they scrambled up the crumbling cliff face onto the Undercliff. But they had great difficulty making their way across the fields because they were now "intersected by chasms."

At first light the full extent of the great landslide became apparent to the homeless cottagers. A "tremendous chasm extending three quarters of a mile from east to west and varying in breadth from two hundred to four hundred feet" had opened along the landward side of the Undercliff. As the chasm yawned open, a fifty-acre parcel of land slipped toward the sea, sinking some fifty feet below its former level in the process. The displaced land became so broken and transected that it had "the appearance of castles, towers, and pinnacles."

News quickly spread through the awakening town of Lyme, and when word reached the Bucklands they hurried off along the shore to inspect the site. The beach was usually deserted at that time of the year, but a steady stream of townsfolk were making their way toward the west, turning the geological wonder into a Boxing Day excursion. Many of the locals had witnessed land-

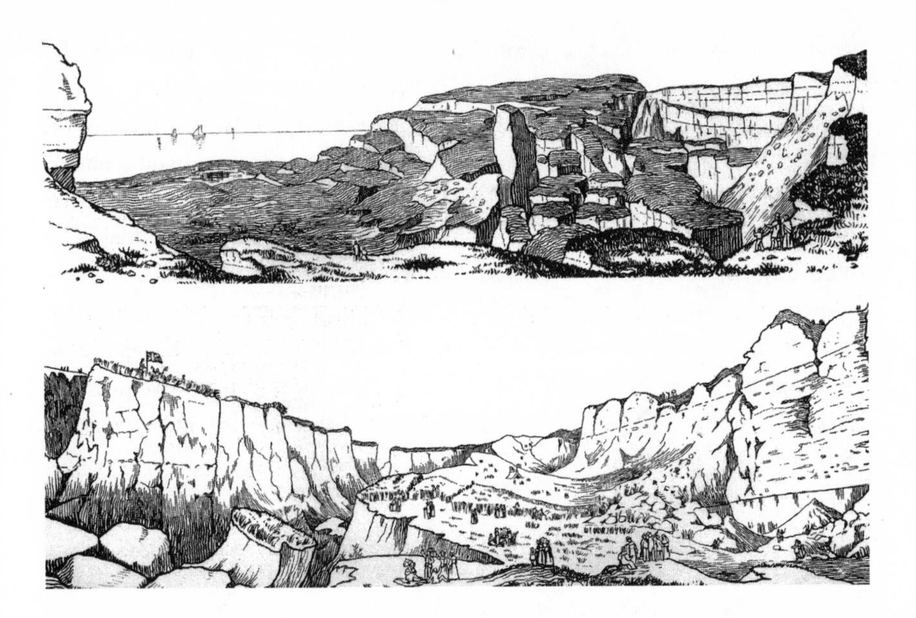

The great landslide of 1839, seen looking toward the sea (top) and toward the land. The event became a Boxing Day spectacle, with flag-waving crowds lining the tops of the cliffs.

slides before, but this was quite unlike any other. A sense of excited anticipation was in the air, lending a carnival atmosphere to the occasion.

Although everyone expected to see something quite spectacular, nobody was prepared for the incredible sight that greeted their eyes. It was as if some gigantic hands had torn a whole section of the Undercliff away from the land and pushed it toward the sea. The same hands had also thrown up a barrier reef just offshore, forming a lagoon about twenty-five feet in depth. Surely no natural forces could visit such a transformation on the land, and all in one night? Some of the townsfolk probably acknowledged God's hand—perhaps even the Devil's—in the deed.

The Bucklands spent several hours at the site. The professor searched for clues to the cause of the geological catastrophe, and Mrs. Buckland employed her artistic talents to capture the surreal scene in a series of detailed drawings.

Buckland correctly attributed the cause of the slippage to the unusually high rainfall of the previous several months. The porous chalk cliffs along this stretch of the coast rest on a thick layer of weakly cemented sandstone, underlain by an impervious layer of clay. As water percolates through the chalk it becomes trapped by the clay, preventing it from draining freely. This causes the sandstone to become saturated with water, converting it into a semifluid slurry that can eventually become too weak to support the overlying mass of rock. When this happened in 1839 the Undercliff slumped, and a yawning chasm opened up as a large part of the Undercliff slipped toward the sea. The reef that rose up just offshore represented the material displaced by the slumping. The soft sandstone reef was soon eroded and breached by the sea, and eventually disappeared. Conybeare joined Buckland at the scene, and the two spent some time discussing the catastrophic event. Their discussion of the inexorable movement of rocks might well have led to a discussion of the movements of glaciers, a subject that had been on Buckland's mind since visiting Switzerland the previous year.

Buckland first met Louis Agassiz in 1834 when the young naturalist was traveling to England. A protégé of Cuvier, Agassiz had come to England specifically to study fossil fishes. Buckland made all the necessary arrangements and accompanied him on an excursion around the country to visit all public and private collections housing fossil fishes. Buckland arranged to borrow 2,000 of these specimens and assembled them at the Geological Society for Agassiz to study. Buckland was also instrumental in getting him a grant of £100 to help defray his expenses. Agassiz was deeply touched by the Bucklands' generosity and affection, and they became lifelong friends.

Buckland probably saw a great deal of himself in the charismatic young fish fossilist. Both men had a contagious passion for natural history, which sometimes blinded them to the practicalities of life, like the need to wash out the kettle used for boiling bones before using it to brew tea. They shared the same conviction of the immutability of species and in God's role in bringing new species into being after former species had been extinguished by great catastrophes, as in the Noachian flood.

Two years after his visit with Buckland, in the summer of 1836, Agassiz received an invitation from Jean de Charpentier, a Swiss geologist of some note, to spend a vacation at his home. De Charpentier lived in the Rhône Valley amid orchards and vineyards in one of the most delightful parts of Switzerland. The idea of taking a holiday with his young wife and their eighteen-month-old son was most appealing to Agassiz, and he happily accepted the invitation. His

A Swiss glacier (at Grindewald), from an 1820 travel book.

congenial host took him on long walks during the day. Evenings were spent in the company of de Charpentier, his German wife, and a convivial group of other guests. They enjoyed good food, stimulating conversation, and fine wine from their host's well-stocked cellar.

De Charpentier was particularly interested in glaciers. Two years earlier he had presented a provocative paper to the Helvetic Society of Naturalists, explaining the likely origin of erratic blocks. These boulders, so named for their seemingly random distribution, were first noticed at the foot of the Alps, sometimes at considerable heights. They are occasionally found in unusual locations, as when precariously perched atop other rocks. The origin of these enigmatic rocks, which can be of immense size, defied explanation.

Lyell discussed erratics in the third volume of *Principles of Geology.* Drawing from the observations of polar explorers, who had seen rocks carried out to sea attached to icebergs, he attributed erratics to that cause. De Charpentier disagreed. He argued that erratics had been transported by glaciers, as they inched their way slowly down a mountainside. De Charpentier did not claim originality for this concept: A mountaineer had shared this view with him in 1815, based on his personal observations.

De Charpentier's presentation on glaciers had been met with much skepticism in 1834, with Agassiz among his more vocal critics. Agassiz had not changed his mind in the interim, and was therefore resistant to his host's glacial arguments when the vacation began. However, when they clambered over rocks together, looking at glaciers firsthand, he quickly became a convert.

The most compelling evidence for the glacial origin of erratics is that their bottom surfaces are polished and scored by parallel scratch marks, with corresponding marks on the rocks on which they lie. Similar patterns may also be seen on the exposed surfaces of rocks embedded in glaciers and on their contact surfaces with the valley walls. The writing on the rocks seemed unequivocal.

Agassiz gave a descriptive account of the evidence for glaciation the year after his Rhône Valley vacation, at a meeting of naturalists in Neuchâtel. During his presentation he introduced the idea of a prehistoric Ice Age, when much of the Northern Hemisphere was covered by a massive sheet of ice. Such a notion struck many in his audience as completely implausible, and his presentation was not well received. He also lost credibility among some of his former advocates. But a few among his listeners were won over. With the passage of time, the idea slowly took root and flourished, at least in Europe.

We know today that much of the Northern Hemisphere and the southern parts of the Southern Hemisphere were extensively glaciated—several times over. Glaciation was responsible for significant changes in the Earth's crust, including carving valleys and depositing gravel beds. There were four extremely cold periods during the Ice Age, when the glaciers spread farther south, in the Northern Hemisphere. These were interspersed with warmer, interglacial periods, when the ice retreated northward. The entire period is referred to as the *Pleistocene,* a term coined by Lyell. Radiometric dating places the start of the Pleistocene at 1.6 million years ago, ending with the retreat of the glaciers some 10,000 years ago.

Agassiz gave a much better argument for glaciation at a meeting in Germany in the fall of 1838. This was attended by Buckland and his wife. After the meeting the Bucklands accompanied their good friend Agassiz to Neuchâtel, his home, before continuing on their European tour. On their return to England Mrs. Buckland wrote and told Agassiz that "We have made a good tour of the Oberland . . . and have seen glaciers, etc., but Dr. Buckland is as far as ever from agreeing with you." Although Agassiz initially had been unable to convince him of the wide-scale effects of glaciers, Buckland now seemed on the brink of changing his mind. As he had informed Agassiz, he had recently seen score marks on rocks in Scotland and England just like those he had seen in Switzerland that were caused by glaciers.

Agassiz visited England the year after the great landslide, attending the 1840 meeting of the British Association for the Advancement of Science, held in Glasgow. The previous year the meeting had been held in Birmingham, and Buckland had given a booming oration in Dudley Cavern. An audience of over 1,000 people had thronged the large gallery, specially illuminated for the occasion with hundreds of flares. Agassiz had a much smaller audience for his presentation at the Glasgow meeting, and his paper on the evidence for glaciation did not convince many people: "Agassiz gave a great field-day on Glaciers," Murchison wrote after the meeting, "and I think we shall end in having a compromise between himself and us of the floating icebergs! I spoke against the general application of his theory."

At the end of the meeting Buckland readily agreed to accompany Agassiz on an excursion to Scotland to search for evidence of glaciation. Murchison decided to go along too. The three companions returned from their Scottish excursion several days later. Murchison was as unconvinced as ever of the likelihood that glaciers could carve and reshape the landscape, but Buckland was fully persuaded. Like most converts he became an ardent proponent of the new cause. He and Agassiz presented separate papers on the evidence for glaciation at the next meeting of the Geological Society. Writing to his friend De la Beche in Wales, just before the meeting, Buckland urged him to attend:

> & give us your opinion about the Glaciers in Scotland on which Agassiz had a paper . . . and I . . . all in favour of the Ice. . . . which in 1/2 an Hour, I w[oul]d. undertake to convince you is equally applicable to the mountain Regions of Wales. Pray come up to the next meeting or you will be behind hand with the March of the Glaciers.

Buckland also tried to win over his close friend Conybeare, but he would not be persuaded. During a winter trip to Lyme Regis, Conybeare wrote to Buckland:

> Though sadly frost bitten at this moment, I don't quite believe in the former Geological Supremacy of the Frost King. I am afraid I see reasons to prove the Universal prevalence of Glaciers *physically impossible*. . . . as to the evidence you *see a few scratches on* the face of a rock & a heap of gravel at its base, & then by an argument *per saltem* get at yr. Q.E.D.

In another letter, Conybeare expressed the view that "All Glacial agency appear too absurd to enter into any ones mind . . . ," concluding that "The Glacial Theory has always appeared to me a glorious example of hasty unphilosophical & entirely insufficient induction."

Buckland, one of the last apologists for the diluvial theory, had seen the ranks of its supporters diminish over the years, with only a few diehards, like Conybeare and Murchison, left to continue the battle. His own conversion to glaciation, after more than two decades of proselytizing the geological evidence for the biblical deluge, may appear as a radical shift in position, but it was probably not. His faith in God, and in the Scriptures, was unchanged, and he still believed in the Noachian flood. All that had really changed was his realization that the flood may not have left an indelible mark on the rocks. Even Conybeare, his wise council on theoretical issues, had raised the question, some long time before, "whether the diluvial traces we still observe geologically, be vestiges of the Mosaic deluge, or whether that convulsion were too transient . . . ." And Buckland's acceptance of glaciation was not a denial of the role played by the Creator. As Agassiz put it, glaciation was "God's great plough."

Lyell was initially convinced by Buckland's arguments, and gave a paper on the geological evidence for glaciation in Scotland, at the Geological Society. But he later abandoned glaciation in favor of rivers as the agents for geological change, a concept Mantell agreed with.

The young sciences of geology and paleontology were maturing, albeit slowly, and the almost universal abandonment of the diluvial theory was a sign that religion's grip on science was weakening. Buckland, in his ecclesiastical enclave at Oxford, was one of the last converts to scientific rationality, at least as far as the Noachian flood was concerned. There was still a long way to go in the separation of science from religion, but the road ahead was beginning to be paved for Darwin.

At the same time the last metaphorical musket shots were being loosed off at the Geological Society in London, real hostilities broke out in Somerset. At the center of the conflict, known locally as the "Sharpham Wars," was the flamboyant Thomas Hawkins.

Sometime after the incident over the strawberries, Hawkins got into a property dispute with Samuel Laver, claiming rights to certain of his farm stock. Hawkins was so convinced he was the wronged party that he attempted to have Laver—the lawful tenant of Sharpham Park—evicted from the property. He was unsuccessful in his bid, but did manage to gain some local support for his dispute with Laver.

As a former resident of Glastonbury, Hawkins ingratiated himself with the townsfolk, no doubt exploiting the historic rivalry between his hometown and the neighboring villages of Walton and Ashcott. Using his natural talent for fomenting trouble, Hawkins polarized the neighborhood, building up such a loyal following among the residents of Glastonbury that they often got into disputes with his detractors in the nearby villages. The clashes between the rival factions became overtly hostile, and full-scale rioting broke out in the villages of Street and Walton one summer Saturday. A number of arrests were made, and Hawkins pressed charges against some of the people taken into custody, presumably for some perceived damages to his reputation.

The following Monday Hawkins set off for Glastonbury to pursue his case against the detainees, accompanied by his solicitor and a servant. But as their carriage approached Walton it was surrounded by an angry mob sympathetic to the detainees. They brought the carriage to a halt, forced Hawkins out, and unceremoniously marched him away to a nearby inn. And there they held him, under duress, and extracted a written statement from him that he would drop all charges against the detainees. He was then allowed to continue on his journey to Glastonbury with his solicitor and servant, followed by a large contingent of the mob.

Although the rioters were being detained at Glastonbury, the case against them was to be heard at the court in Wells, nine miles away. The prisoners were therefore loaded aboard a wagon and driven to Wells, under a guard of special constables. Hawkins and his party tried to leave for Wells too, but the mob pushed back his solicitor and servant, and only he was allowed to proceed to court.

The counsel for the prosecution, one Mr. G. Chitty, acted on behalf of Hawkins. He contended that the statement his client wrote for his captors was obtained under threat of violence, and it was accordingly dismissed. He then told the court that all his client sought was the protection of his property, and called upon the magistrate to interpose in the dispute. After a long examination of the circumstances that led up to the riot of the previous Saturday, the magistrate made his ruling. Four constables were to be posted at Sharpham Park to keep the peace, two for Hawkins and two for Laver. The men arrested for rioting were bound over to keep the peace, and released.

The unfortunate Samuel Laver died later that summer, and the trustees of his estate served Hawkins notice to vacate the premises. Hawkins refused to comply, so the trustees broke into his part of the house in an unsuccessful attempt to evict him. Hawkins immediately took legal action against them for alleged damages to his property, serving them with upward of ten writs. The alleged damages included the replacement of a case of wine knocked over by a dog and the death of a valuable horse that caught a chill after a stable door had been left open. The horse in question was still alive and well at the time the writs were served.

Hawkins defiantly refused to vacate the premises, and hired a group of thugs to ward off any attempts to evict him. One of their functions was to seize anyone who approached the house too closely. Offenders were forced inside, and after being lectured by Hawkins, they were horsewhipped and ejected. Such blatant acts of violence did not pass without retaliation, and Sharpham Park became a venue for regular battles between Hawkins's supporters and his detractors. The noise of the disturbances was often so loud at night that it could be heard in the adjacent villages of Walton and Street.

On one particular occasion Hawkins was returning home from Bridgewater when an angry mob from Walton surrounded his carriage and dragged him out. They were all set to do grievous harm to his person when the rector made a timely appearance on the scene. By dint of persuasion he managed to free Hawkins from their clutches, and took him into the sanctuary of the rectory. The rector then used the same powers of reasoning to persuade Hawkins to seek an arbitrated settlement to his dispute with Laver's trustees: Nothing short of that would end the Sharpham Wars. Hawkins, no doubt sobered by his rough treatment at the hands of the mob, acquiesced to the rector's entreaty.

Hawkins had his day in court, but the decision went against him and he was ordered to vacate Sharpham Park. He claimed the ruling was unfair because the barrister who acted as the arbitrator was a cousin of the rector's wife. But his arguments held no sway with the court, and the eviction order was duly carried out. Having arrived like a lion, Hawkins appears to have left the neighborhood like a lamb. But he left the trustees with a crippling legacy of expenses and damages, amounting to £5,000 or £6,000. This brought personal ruin to each one of them. Ironically, one of Laver's trustees was the father of the young man Hawkins had struck during the strawberry incident. The worry over his financial ruin was said to have contributed to his premature death, and the debt passed to his son. He subsequently became bankrupt too, and emigrated to New Zealand. The effects of Hawkins's troublemaking had far-reaching consequences. But Somerset, at least, was well rid of him.

14

Of Dinosaurs and Species

Monday, August 2, 1841, was a rather dreary day in London, with overcast skies and the threat of rain. Gideon Mantell was at home in Clapham with Reginald, the youngest of his children. Reginald had been staying with him over the weekend, but Mantell had to take him back to school in Hanwell, a ten-mile journey by train. Mantell had recently discovered the wonders of rail travel. Just two weeks before he had taken a one-day excursion to the West Country on Brunel's Great Western Railroad. Starting at ten, he had enjoyed a whole day's fossiling and rambling around his early childhood haunts. He returned home, laden with fossils, at eleven at night, having had breakfast in Clapham, and lunch and supper in Wiltshire.

While Mantell was delighting in the company of his son most of his geological circle were attending the annual meeting of the British Association, held that year in Plymouth. Mantell might have attended—he was still wedded to his fossils—but probably felt a greater need to spend more time with his family. He had even normalized relationships with his daughter Ellen, now twenty-three. He would be driving her to Hanwell the following weekend to visit her brother. But Mantell saw nothing of his wife.

The Plymouth meeting was an historically important one for Owen, but he probably did not realize it at the time. With De la Beche—now Sir Henry—in the chair, he was about to give his second report on the British fossil reptiles. Those in the audience who attended the reading of Owen's first report, two years before, probably remembered that he spoke for two and a half hours, and this promised to be another marathon lecture. Having made his formal introduction, Sir Henry sat down, and Owen took his place at the podium.

Owen's restoration of *Megalosaurus,* showing its erect posture on vertical legs. The head looks too crocodilian, and modern restorations depict a bipedal posture.

Owen cut an impressive figure—tall, self-assured, with a commanding voice and penetrating eyes. Cuvier had been dead for almost a decade, and Owen was now widely recognized as the world authority on vertebrate anatomy, both of the living and the dead. After some brief remarks on the reptiles that were the subject of his paper, he paid his respects to the pioneers in the field: Buckland and Mantell. He talked about plesiosaurs in brief, then crocodiles in depth, going into some detail on their anatomy. Anyone doubting the duration of his presentation had only to glance at the great thickness of his notes to fear the worst.

Owen directed his audience's attention to *Iguanodon, Megalosaurus,* and *Hylaeosaurus.* These three "creatures far surpassing in size the largest of existing reptiles [formed] a distinct tribe or sub-order of Saurian Reptiles. . . . " Owen did not actually use the term "dinosaur" in his Plymouth address, as historian Hugh Torrens has shown. However, in his written account published the following year, he proposed "the name of *Dinosauria,*" meaning "fearfully great lizard."

Among their characteristic features was a "large sacrum composed of five anchylosed [fused together] vertebrae. . . . " No other reptiles had more than two sacral vertebrae. The great behemoths therefore had their hips attached to an extensive bony union of their backbone, "as in land mammals." Their legs and feet were similarly constructed as massive supporting structures. In discussing *Megalosaurus,* Owen told his audience that the hind part of the body was raised "high above the ground upon long and strong hind-legs [which] must have been different from that of any existing Saurians. . . ." Owen made similar comments for *Iguanodon,* which he contrasted with the modern iguana. As in other lizards, the femora of the iguana extended horizontally from the body, "which is rather slung upon than supported by those bones. . . ." However, in *Iguanodon* the legs "must have sustained the weight of the body in a manner more nearly resembling that in the pachydermal Mammalia."

Owen clearly recognized that most of the dinosaur's weight was carried by the hind legs. These were kept vertically beneath the body, not splayed out at the sides as they are in modern reptiles. Remarkably, Owen reached this conclusion on the evidence of only three genera, each one of which was very incompletely known. There were other genera too, including *Cetiosaurus* (a sauropod), and *Thecodontosaurus* (a prosauropod), which he mentioned in his report, but he did not recognize them as dinosaurs at the time (he thought *Cetiosaurus* was a crocodile and *Thecodontosaurus* a lizard). Today we have several hundred genera of dinosaurs, many represented by complete skeletons. Their confirmation of Owen's original conclusions is a fitting testament to his anatomical insights.

Owen was most respectful of Mantell throughout his presentation. At one point he referred to "the kindness of Dr. Mantell" in making certain specimens available to him. However, some of his comments on *Iguanodon* would have rankled Mantell had he been in the audience. Referring to the great size and strength of its hind limbs, Owen commented that "One can scarcely suppress a feeling of surprise, that this striking characteristic of the *Iguanodon,* in common with other *Dinosauria,* should have been, hitherto, overlooked. . . ." He was also very critical of Mantell's length estimates in his written report, noting that "it is very obvious that the exaggerated resemblances of the *Iguanodon* to the Iguana have misled Palaeontologists." Scaling up a modern iguana, as Mantell had done, so that its femur was the same length as that of *Iguanodon* gave an estimated body length of seventy-five feet (23 meters), which was far too great. Owen, in contrast, based his body length estimate on vertebral lengths. This, of course, required estimating how many vertebrae were present because the most complete specimen of *Iguanodon* then available, the jumbled remains of the hindquarters (the Maidstone specimen), was far from com-

Owen's restoration of *Iguanodon* showing vertical legs—fundamentally different from the sprawling posture depicted by Mantell.

plete. Regardless of this, Owen's estimate of twenty-eight feet (8.6 meters) was remarkably accurate, as we know today from our knowledge of complete skeletons.

Owen had similar criticisms of Buckland's length estimates for *Megalosaurus*, reporting that the sixty- to seventy-feet (18–21 meters) approximations were based on "the fallacy of concluding that the locomotive extremities bore the same proportion . . . as they do in the small modern land lizards." Again, basing his length estimate on the size of the vertebrae, Owen arrived at a body length of thirty feet (9.2 meters). *Megalosaurus* continues to be as poorly known today as it was in Owen's time, and we cannot measure its actual length. However, based on comparisons with similar genera, Owen's estimate is probably close to the truth.

Owen's scientific objectivity coexisted with his belief in the creative hand of God. This is apparent from two allusions he made to a creative being. The first was in reference to the organization of the vertebrae in the sacral region, "which Creative Wisdom adopted to give due strength. . . . " The second was made during his discussion of the way the teeth of *Iguanodon* were adapted for their grinding function. He observed that the outer layer of the tooth was formed of enamel, which, being harder wearing than the dentine forming the rest of the tooth, retained its cutting edge. Quoting from Buckland's *Bridgewater Treatise,* he considered this adaptation had "resulted from design and high intelligence."

Owen's discussion of the dinosaurs occupied only about a quarter of his report, and was an uncommonly low-key part of his presentation. This might seem odd to us today, considering their current popularity, but our present preoccupation with dinosaurs dates back less than three decades. That is not to say that Owen, or his audience, were not intrigued and interested in the announcement of a new group of extinct reptiles. It is just that the occasion was not marked by the kind of attention that a similar revelation would attract today. Most of the rest of his talk was devoted to an in-depth account of the other reptilian groups, which included lizards, pterosaurs, turtles, and snakes.

His entire review took a little over two hours, by which time even his most ardent followers must have been getting a little fidgety. But he was not yet finished because he still had to give a rather lengthy summary of his findings.

We can picture him glancing up from his lecture notes—the daunting pile much reduced by his previous labors—and posing a series of rhetorical questions for his audience. Did the "numerous, strange, and gigantic" reptiles live in the place in which they were unearthed? Or were they "the relics rather of antediluvian creatures . . . washed from latitudes suitable to their existence . . . ?" And were they actually extinct, or might they "still remain to be discovered in those warmer regions . . . ? Such questions," he offered, are "most likely to suggest themselves to those who are not conversant with the truths of Geology. . . . " In case any of his audience were in any doubt as to the correct answers to these questions, he answered the first in the affirmative and the other two in the negative.

He then got down to the main issue, which was whether the successive genera and species of fossil reptiles supported "the transmutation of species, by a march of development occasioning a progressive ascent in the organic scale. . . . " He told his audience that a superficial survey of the fossil record might appear to support just such a notion. But he warned them that "of no stream of science is it more necessary, than of Palaeontology, to 'drink deep or taste not.'"

Owen saw no evidence for a progressive change from the simple to the more complex through time, and supported his case by reference to the ichthyosaurs. He told his audience that it was now possible to trace their long history through an immense series of strata, but that "the very species, which made its first abrupt appearance" in the Early Jurassic, "maintains its characters unchanged and recognizable" in the Late Cretaceous. As far as he was concerned there was "no evidence whatever that one species has succeeded or been the result of the transmutation of former species."

Broadening his scope to encompass all the reptiles reviewed in his report, he reasoned that if modern reptiles were the result of a "progressive development and transmutation of former species," each group should now be at its highest level of organization. This was clearly not the case, and he illustrated his point by reference to the dinosaurs. No living reptiles had such complicated thecodont (set in sockets) teeth, or such a well-developed sacrum and hind legs. He continued this line of reasoning to argue that if the dinosaurs were "on the march of development to a higher type, the *Megalosaurus* ought to have given origin to the carnivorous mammalia, and the herbivorous [mammals] should have been derived from the *Iguanodon.*"

Owen made the point that the time period when reptiles were at their peak, both in their level of organization and in their numbers, was long past, and that they have been in decline ever since. The reptiles had been superseded by the mammals, and although new reptilian species had "constantly succeeded each other . . . the change has been, upon the whole, from the complicated to the simple."

Owen's lengthy summary reiterated his opposition to the idea that species were transmutable. It also affirmed that, like Lyell, he did not see any progression in the fossil record toward "higher" animals. As far as Owen was concerned, the dinosaurs represented the very pinnacle of reptilian organization, and modern reptiles paled in comparison.

The conclusions he reached were quite reasonable, based solely on the evidence he presented for the British fossil reptiles. He was correct in saying that the characters of the ichthyosaurs he studied had remained unchanged. However, he had relatively few specimens at his disposal, and almost all of them were from the earliest Jurassic. Specimens from the Cretaceous were fragmentary—mostly isolated teeth and vertebrae—and material from the early Triassic were unknown. We have considerably more ichthyosaurs to study today, including thousands of complete, or near complete, skeletons. They range in age from the Early Triassic to the Late Cretaceous, and show large-scale evolutionary changes during their 150-million-year history. For example, the earliest ichthyosaurs had relatively long and flexible bodies that lacked a

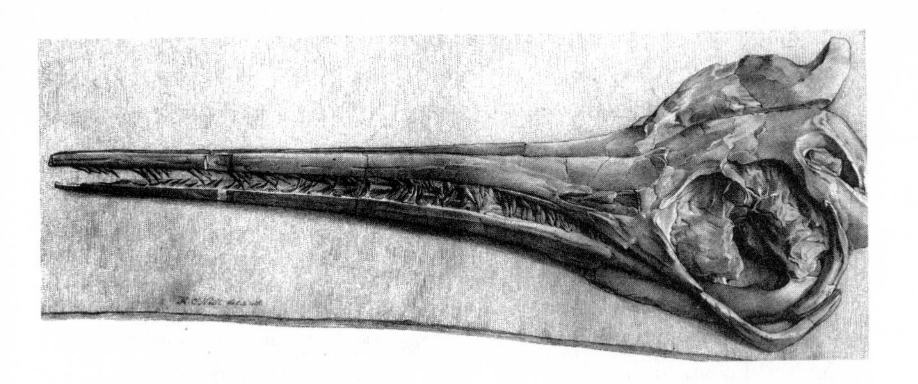

The long-snouted ichthyosaur *Ichthyosaurus tenuirostris*, now referred to the genus *Leptonectes*. (Skull length 21 inches; 54 cm)

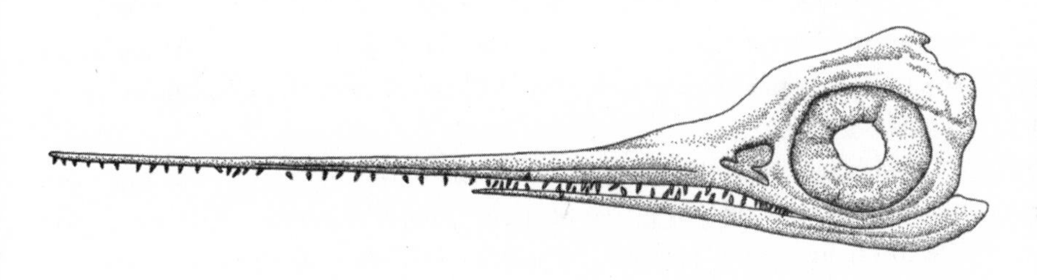

The swordfishlike ichthyosaur *Eurhinosaurus*. (Skull length 5 feet; 1.5 m)

tail bend. Only later did they evolve the crescent-shaped tail that characterizes the high-performance swimmers of the Jurassic.

Owen discussed the different types of ichthyosaurs, like the long-snouted *Ichthyosaurus tenuirostris* and the "shorter jawed *Ichthyosaurus communis,*" but saw "no evidence whatever that one species has succeeded or been the result of the transmutation of a former species." It is certainly true that the fossil record is strikingly poor in transitional forms between one type and another, a fact lamented by Darwin, almost two decades later, in his *Origin of Species.* However, transitional fossils—the proverbial "missing links"—do exist.

For example, among the most specialized ichthyosaurs is one called *Eurhinosaurus,* discovered in 1909. The name, which means "true-snouted reptile," was chosen because of its remarkably long rostrum. This extends well beyond the tip of the lower jaw, like the "sword" of the living swordfish.

Eurhinosaurus seemingly appeared without antecedents, and no other ichthyosaurs were even remotely like it. Then a remarkable discovery was made in 1986, along a desolate stretch of the Somerset coast. Here was an ichthyosaur with a partially developed sword. The new ichthyosaur, which I named *Excalibosaurus* after the sword of Arthurian legend, appears to be intermediate between the long-snouted species that Owen mentioned and *Eurhinosaurus.*

A much more celebrated example of a transitional fossil is *Archaeopteryx,* the earliest bird, discovered in 1861 just two years after publication of the *Origin of Species.* Possessing a mixture of avian and reptilian features, *Archaeopteryx* formed a perfect link between reptiles and birds. Thomas Huxley studied *Archaeopteryx* and used it to support the case for evolution. Owen also studied *Archaeopteryx,* but he saw it as nothing more than a bird, albeit a rather unusual one with its long bony tail and clawed wings. He flatly rejected these peculiar features as being reptilian, thereby denying the transitional nature of *Archaeopteryx.* His scientific mind, like Buckland's before him, had already been made up.

Owen's argument that there was no evidence for the transformation of one species into another, no "march of development," whereby *Megalosaurus* gave rise to carnivorous mammals, was a straw man. But, in fairness to Owen, it was not a mannequin of his own making. This is because the popular concept of species transmutation—evolution—in 1841 was that one species was literally transformed into another, the way an infant is transformed into an adult. The world had to wait another decade for Darwin to present a more subtle transition, whereby a mother species gave rise to a daughter species, over countless generations of gradual change.

Darwin's theory of evolution is elegant in its simplicity. Offspring are similar, but not identical to, their parents. Species produce far more offspring than can possibly survive. Because the offspring are not identical, it follows that some individuals have features that give them an advantage over others. These advantages give the individual a better chance of surviving, and therefore of leaving more offspring. Because the offspring of the advantaged individuals inherit some of the favorable features of their parents, they too have an improved chance of survival. Darwin termed this process, the selection of advantaged individuals, *natural selection.* The operation of natural selection, over numerous generations, results in the accumulation of sufficient change to give rise to a new species.

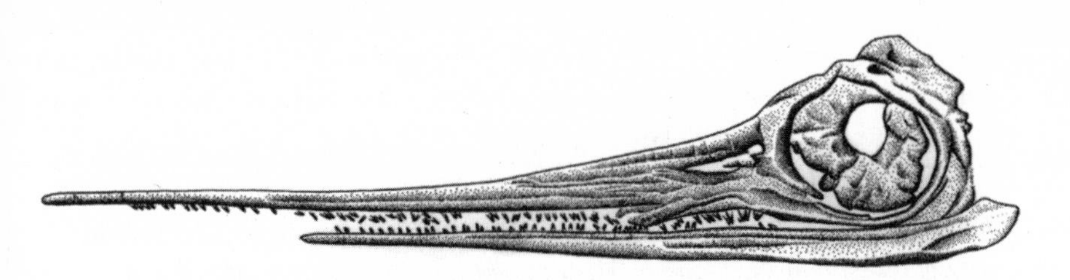

Excalibosaurus, with its partially developed sword, appears to be intermediate between long-snouted ichthyosaurs like *L. tenuirostris* and *Eurhinosaurus*. (Skull length 30 inches; 78 cm)

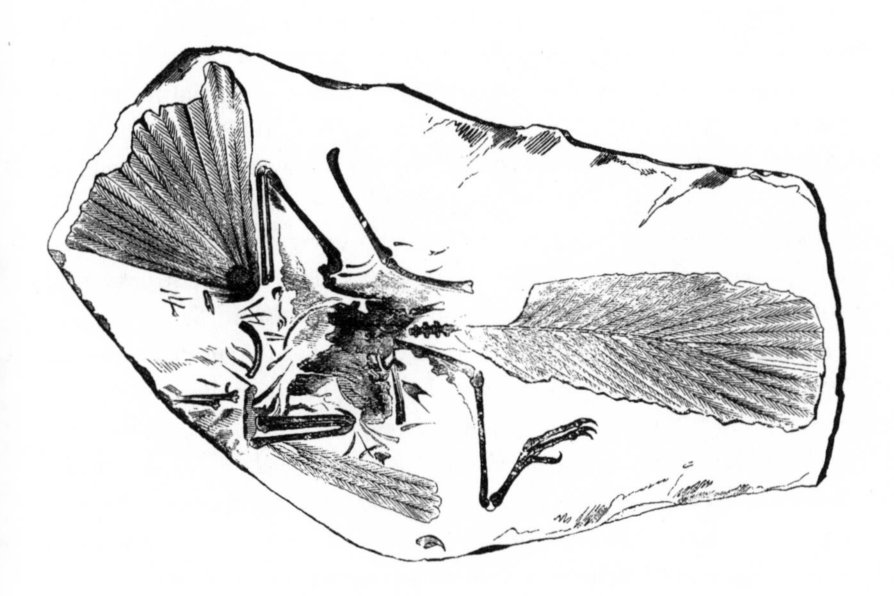

The earliest bird, *Archaeopteryx,* has a mixture of avian and reptilian features.

If Owen saw no "march of development" in the fossil record it was because he was being selective. Owen knew as well as anybody that the earliest vertebrates were the fishes, and that animals without backbones appeared in the fossil record before the appearance of vertebrates. But he chose to ignore this as evidence of a "progressive ascent in the organic scale." As far as Owen was concerned the fossil record simply did not support progression. Darwin, on the other hand, saw an increasing diversity and complexity of living things, and this accorded with his view of how new species came into existence, over long periods of time, through natural selection.

Owen would write a scathing and anonymous review of the *Origin of Species* in the *Edinburgh Review,* and remained opposed to Darwin for the rest of his long and distinguished career. History has not been kind to him for this, but we should remember Owen's outstanding contributions to anatomy and paleontology. His recognition of the dinosaurs as a distinctive group—his synthesis of what had previously been regarded as a disparate assemblage of scaled-up modern reptiles—was a brilliantly incisive contribution to science. This work is all the more impressive when account is taken of how few dinosaurs were known at that time, and how incomplete they were. And his dinosaurian study was just one among a multitude of contributions. Like many other biologists, I have often used Owen's publications as a starting point for my own research, from his studies on ichthyosaurian anatomy to his dissection of the leg muscles of the kiwi. Owen was right about anatomy most of the time, and the durability of his studies is a fitting tribute to his outstanding abilities. We should also remember that Owen's negativism toward species transmutation and the "march of development" in the fossil record would have struck a resonant chord with most of his audience on that historic day in 1842.

Buckland was the first on his feet at the end of Owen's tour de force of the British fossil reptiles. He acknowledged Owen's labors in the most complimentary of terms, and enthusiastically expressed his appreciation for the interest he had stirred in the audience. When Buckland made the comment that Owen was a worthy successor to Cuvier, there was a spontaneous burst of applause throughout the entire room. Owen looked down upon the adulating throng from his lofty perch—he might have been on Mount Olympus.

While Buckland applauded, and Owen smiled, and Mantell enjoyed his son, Anning perambulated the foreshore in search of fossils, as usual. Owen had been unstinting in his acknowledgments of Buckland and Mantell's contributions to the knowledge of dinosaurs, but, when it came to the ichthyosaurs, she received not a single mention. Anning would eventually hear the highlights of the Plymouth meeting, probably from Buckland on his next visit.

These were Owen's daffodil days, when everything was fresh and green and blooming. He was the anointed one, the heir to Cuvier's throne, and was much in demand for his professional opinions. Darwin frequently sought his views, and the Owens saw a good deal of Darwin and his new wife at this time. The Darwins lived in a small house on Gower Street, in central London, which seemed much to their liking: "We are living a life of extreme quietness. . . . and if one is quiet in London, there is nothing like its quietness—there is a grandeur about its smoky fogs, and the dull distant sounds of cabs and coaches; in fact . . . I am becoming a thorough-paced Cockney. . . ."

This was a departure from his earlier experiences of London, which he said was "a vile and smoky place. . . ." Perhaps married life had occasioned the change. However, within only two years they were planning to move to the countryside, where Darwin longed to "be settled in pure air, out of all the dirt, noise vice & misery of [London]." His disillusionment with London probably had much to do with the poor health that had troubled him since his return to England. He was "about six feet in height," but, by the spring of 1840, he weighed only 148 pounds (67 kg), having lost 10 pounds since the previous fall. In an attempt to restore his health he took frequent trips to the countryside, often to the family home in Shrewsbury. "The country at first acted like magic on me. . . . I am . . . a good deal stronger than when in London, but . . . the Doctors tell me, it will be some years, before my constitution will recover itself. . . ."

His illness, probably contracted during his travels, drained him of energy, causing him to curtail most of his activities, such as attending Geological Society meetings. "I am forced to live . . . very quietly and am able to see scarcely anybody. . . ." Owen was one of the few people he saw regularly during this period.

Owen and Darwin got along very well during these years, each perhaps appreciating the other one's radically different background and experience. Owen was a laboratory man, who had gleaned most of his knowledge of animals through his skill with the scalpel. Darwin, on the other hand, was a naturalist and collector, who learned his trade through his meticulous observations of the living world. Each man gained from the fresh perspectives the other brought to their zoological discussions. But Darwin always had to be on guard not to let slip any mention of his clandestine work on the transmutation of species. If Owen had ever found out it would likely have ended their friendship, but Owen unwittingly continued to supply Darwin with information useful to his covert cause.

Leading a double life must have taken its toll on Darwin, a warm-hearted soul, whose natural inclination was to be open with others. Fortunately he could confide his secret work to his small circle of confidants. "I continue to collect all kinds of facts, about 'Varieties & Species' for my some-day work . . . ," he wrote to his cousin, William Fox. He then went on to express his interest in any animals that Fox might donate to the cause: "the smallest contributions, thankfully accepted . . . if your half-bred African Cat should die . . . I should be very much obliged, for its carcase [*sic*] sent up in a little hamper for skeleton . . . it . . . will be more acceptable than the finest haunch of Venison. . . ."

While Darwin worked on the species problem and began to amass the tremendous body of evidence that his theory would depend upon so heavily, Owen took his first major step up the social ladder. This came toward the end of 1842, in a letter from the prime minister, Sir Robert Peel. Sir Robert, who had taken a particular interest in Owen's flourishing career, informed him that he had been awarded an annual pension of £200 from the Civil List. The prime minister stressed that the acceptance of the pension would in no way fetter his independence, and that it had nothing to do with his political opinions, of which he was totally unaware. His only objective was for the Crown "to encourage that devotion to science for which you are so eminently distinguished." Buckland, who was an old acquaintance of Sir Robert's and a familiar face at Peel's country home, had been instrumental in securing the pension for Owen.

The prime minister's letter caused a great deal of excitement in the Owen household. Aside from being a most prestigious honor, the pension would provide a significant source of additional income. As soon as Owen had digested the contents of the letter, he pulled on his boots and set off for Broderip's house. His good friend was just about to retire, but soon hurried on his dressing-gown and they sat down to work out an appropriate response. Before Owen set off for home again, Broderip poured some sherry on the ground as a libation.

Congratulatory letters trickled, if not flooded, into the Royal College of Surgeons, including one from William Whewell, the Master of Trinity College, Cambridge. He hoped the "well deserved honour will have its weight in protecting you from the molestation of those who might otherwise not acknowledge your value." Whewell, like many others, was well aware that Owen had his detractors.

Buckland, with his penchant for appearing unexpectedly, materialized at Owen's apartment a few weeks later, suggesting they call on Sir Robert Peel. During their short conversation, Sir Robert asked Owen when he might visit the Hunterian Museum. The following Saturday was agreed upon, and the prime minister duly arrived, accompanied by Buckland. He spent an absorbing two hours examining the collection, being as interested in the specimens as he was in the brilliant young professor conducting the tour. Peel, it seemed, had further plans for Owen.

The following year Owen was invited to serve on a Royal Commission to investigate the public health of large towns. He served on this, and other commissions, for the next several years, putting his scientific knowledge to practical use for a number of issues, from sewers and drains to slaughterhouses. Owen took his responsibilities very seriously, setting off early in the mornings to inspect some of the poorest and most dilapidated parts of London. A police officer accompanied him when he visited the more notorious areas, and he would return home "quite distressed at the misery and filth he had witnessed." One morning he set off in a particularly thick industrial fog—a legacy of the coal-burning fires—in a "desperate determination to find his way to Whitechapel." This was the district later made infamous by Jack the Ripper. After some considerable difficulty he found the slaughterhouse he was looking for. He made his inspection of the gory facilities, then groped his way home, arriving back several hours later.

Owen's Royal Commission duties brought him into contact with some of the most influential people in the land, and he was often seen in the company of Peel and his ministers. The prime minister commissioned the society painter, Henry Pickersgill, to capture Owen's likeness on canvas. When the work was finished the portrait was hung in Peel's private gallery of eminent men. Owen struck a handsome pose, looking every bit the suave and debonair gentleman he had become.

Within six months of the completion of the portrait, De la Beche called on Owen with a communication from Sir Robert Peel. Would a knighthood be acceptable to him? After talking the matter over with his wife, Owen decided to decline the honor. "It would not add much to our comfort or respectability," Caroline wrote, but if the time ever came when the Hunterian Museum became "part of a great national museum, then it might all be very well."

It seems remarkable that a man of Owen's ambition should have turned down a knighthood. The real reason probably had to do with political pressures from his college. Many of its most senior members had not been honored by the Crown, so a knighthood for Owen, who was still only forty-one, would have been considered inappropriate.

At about the time Owen declined the knighthood a book was published in England that caused a major sensation. Part of the reason for this was that its author was anonymous, but the main reason for its notoriety was its subject matter. The Reverend Adam Sedgwick, who wrote a scathing ninety-page review of the book, said: "From the bottom of my soul I loathe and detest [it]." The book was *Vestiges of the Natural History of Creation,* and the reason the author went to such lengths to hide his identity—he took his secret to the grave—was that it dealt with the taboo subject of species transmutation. Written in a popular style to reach the widest readership, it was an instant success, and was reprinted three times during its first six months. It would run to ten editions, and sell nearly 24,000 copies. Although there was never any disclosure of the book's author, some people, including Darwin, suspected who it was. Some had even suggested that Darwin himself was the author, but nobody familiar with his erudite writing style could possibly mistake this odd mixture of fact and fantasy with his pen. Darwin thought "the writing and arrangement are certainly admirable, but his geology strikes me as bad, and his zoology far worse."

Vestiges began with an overview of the cosmos. This was followed by a chronological review of the fossil record, where the point was made that there was a progression toward "higher forms of organization." Aside from a few glaring errors, as when reptiles were said to "advance to birds, and thence to mammalia," the facts were essentially correct. (Birds and mammals evolved independently of one another.) The author related this progression in animals to physical changes on the Earth. The present arrangement of the continents, for example, was a prerequisite for the appearance of our own species. The author was in no doubt "That God created animated beings. . . . " It was "a fact so powerfully evidenced . . . that I at once take it for granted." However, he had some problems with the traditional view of how this had been achieved. Simply stated, the Genesis account of the Creation was unsatisfactory and some alternative had to be found to account for the way "the Divine Author proceeded in the organic creation."

The idea that God created every species "personally and specially on every occasion" was dismissed as being "too ridiculous." Instead, the case was made that God created every species as a result of "natural laws which are expressions of his will." Attempting to discover how this was achieved, the author reviewed a number of natural phenomena, selected to show a continuity between nonliving and living systems. These ranged from the fernlike growths of ice seen on winter windows to the supposed instances of spontaneous generation of living beings. The most fanciful of the latter examples included the experiments of a Mr. Crosse, who had supposedly created insects by passing an electric current through a solution of copper nitrate! The author attempted to show how seemingly diverse groups of animals were actually linked. He made his case by citing some outrageously misinformed examples, such as the supposed link between birds and mammals provided by the ostrich, which had "a diaphragm . . . and feathers approaching the nature of hair."

The author held that the key to understanding the way one type of animal became transmutated into another lay in its development: "It is only in recent times that physiologists have observed that each animal passes, in the course of its germinal history, through a series of *permanent forms* of various orders of animals inferior to it in the scale."

The author's emphasis upon "permanent forms" drew attention to his own ignorance. Had he attended Owen's Hunterian lectures he would have known that the various stages an animal passed through during its embryonic development should be equated with *embryonic* stages of other animals, not adult, or "permanent" ones. The author illustrated his point with a glaring gaffe: "Thus . . . an insect . . . is, in the larva state, a true annelid, or worm, the annelida being the lowest in the same class." The author clearly thought that a caterpillar, with its superficial resemblance to an earthworm, *was* a worm. As for our own species, "His organization gradually passes through conditions generally resembling a fish, a reptile, a bird, and the lower mammalia. . . . " New species therefore came into being by the addition of further stages to the embryonic development of an existing one:

> Thus, the production of new forms, as shewn in the . . . geological record, has never been anything more than a new stage of progress in gestation . . . the simplest and most primitive type . . . gave birth to the type next above it . . . and so on to the very highest, the stages of advance being in all cases very small—namely, from one species only to another . . . .

Vestiges, like the curate's egg, was good in parts. It offered an explanation for the progressive appearance of more "advanced" types of organisms in the fossil record, through the operation of the "natural law" of embryonic development, which was itself progressive. He reconciled his views with religious beliefs by emphasizing that the natural law was of God's making. Simply put, embryonic development was God's way of bringing new species into existence. However, it did not provide a mechanism for how such changes took place.

We know today that developmental changes are a key in effecting evolutionary change, and that minute modifications during embryonic development can have far-reaching consequences for the adult. Recognition of the importance of embryology placed *Vestiges* ahead of its time, but this was negated by the gross oversimplification of the mechanism and by the glaring errors woven into the fabric of the book. But since most of the readers of *Vestiges* were blissfully ignorant of the facts themselves, this did not detract from their reading of the work.

The immense popularity of the book was of grave concern to the establishment, who feared for their own positions of authority during these politically troubled times. Sedgwick predicted "ruin and confusion in such a creed," which would "undermine the whole moral and social fabric [bringing] discord and deadly mischief in its train."

Murchison wrote an anxious letter to Owen, imploring him to act: "of all persons in this town, you were the most fit to review the "Vestiges." . . . Now . . . that the book . . . is really taking considerable hold on the public mind, a real *man in armour* is required . . . you would do infinite service to *true* science and sincerely oblige your friends. . . . "

But Owen chose not to act. Remarkably, he was sympathetic toward the book, and this is apparent from the letter he wrote to its anonymous author. He told him he had perused the book:

> with the pleasure and profit that could not fail to be imparted by . . . one who is evidently familiar with the principles of so extensive a range of human knowledge. It is . . . the discovery of the . . . causes [of] the production of organised beings . . . [that] is probably the chief end which the best anatomists and physiologists have in view.

Owen's long letter corrected several of the author's more blatant errors, like his examples of spontaneous generation, which Owen correctly attributed to contamination. His letter displays a remarkable and uncharacteristic tolerance for the anonymous author's transgressions. More important, it reveals Owen's conviction that discovering a mechanism to account for the "production of organised beings," that is, for the origin of species, was a primary objective of anatomists. Such a view may seem difficult to reconcile with his later rejection of Darwin and his *Origin of Species.* But one of the problems Owen had with Darwin's thesis was that it was mechanistic and did not invoke the guiding hand of the Creator. Another problem may have been that it was not Owen's idea.

Speculations on the author's identity were rife at the time *Vestiges* was published, no doubt contributing to its brisk sales. Included among the suspects were Charles Lyell, William Thackeray the novelist, and Prince Albert, Queen Victoria's husband. The real author was Robert Chambers, a self-taught writer and editor of *Chambers's Edinburgh Journal,* a weekly digest for the education of the working classes.

Vestiges was widely read and discussed, which was perhaps its greatest value. By encouraging people to think about the unthinkable—the origins of new species—it helped lay the groundwork for Darwin, just as the fossilists had done with their discoveries. Darwin himself acknowledged this positive aspect of *Vestiges* in the third (1861) edition of the *Origin of Species.*

15

Decline and Fall

They called it "Westminster Fever" because it had not spread beyond the precincts of the old London town. But the physicians called it by its familiar name: typhoid. There had been several deaths, and the city had become mournfully used to the unfamiliar tolling of the abbey bell. There was little doubt from whence the pestilence had come. It had been lurking, unheeded and undisturbed, in a disused drainage duct in Little Dean's Yard. But the genie was released from the bottle when some workmen accidentally broke into the ancient conduit during their renovations of the abbey's drainage system. Buckland, the initiator of the renovations, was superintending the work at the time, and was therefore well aware of the mischief that had been done. Buckland, or Dean Buckland as he was now called, had been making some major and long-overdue changes at Westminster Abbey since his new appointment in 1845. He and two of his daughters contracted typhoid during the outbreak, but all three recovered. The disease eventually ran its course, the abbey bell stopped tolling, and the whole distressing episode became a dark and distant memory.

Some people had been surprised to see Buckland leave Oxford after so many years, but not those acquainted with the ways of that university. Conservatives at Oxford were in the ascendancy, and teachers of science had become the poor church mice. The mice had made overtures to university authorities to invoke compulsory attendance and to establish a separate honors school for the natural sciences. Buckland was invited to join them in their press for reforms, but the dispirited geologist, tired of swimming against the tide, had declined. Accordingly, when his old acquaintance, Sir Robert Peel, offered him the newly vacated position of Dean of Westminster, the sixty-one-year-old professor jumped at the opportunity. The Bucklands still retained their country cottage at Islip, just outside Oxford, and returned each summer during the school vacations.

The menagerie was transported from Oxford in its entirety, but Tiglath Pileser, the bear, was no longer himself. His stuffed remains were mounted in the entrance hall, together with those of Billy the hyena. Jacko, the monkey, lived in the kitchen, tethered to a pole, where he had a plentiful supply of black beetles to eat. The eagle was similarly tethered to a pole in the small Deanery garden.

As the Dean of Westminster, Buckland had charge of the abbey, built during the eleventh century by Edward the Confessor. He was also responsible for the adjoining Westminster School, a seat of learning for young gentlemen. His broad responsibilities encompassed the souls of his parishioners, the minds and well-being of the boys, and the fabric of all the buildings. It was to these buildings, sadly decayed and abused through years of neglect, that he turned his practical talents. He used his geological knowledge to select the stone used for restoring the external fabric of the abbey. And his fund-raising skills were put to good use to find the money to pay the bills. Young Queen Victoria herself was one of his patrons.

The Deanery, the family's living quarters, was a huge, rambling building. There were four cavernous wings connected by large paneled landings and passages, with no fewer than sixteen staircases. The entire building was notorious for its frigid chill, and a blazing fire was kept burning in the antechamber, winter and summer. The servants frequently complained of strange noises at night and of gusts of wind that blew out their candles. On one particularly stormy night a section of the wainscoting came crashing down in one of the passageways, frightening some of the servants out of their wits. The next morning a hidden shaft was discovered behind the wall, leading to a secret room. A similar hiding place was found behind one of the walls in the library.

The gregarious Buckland had a continuous stream of visitors and guests calling at the Deanery. Michael Faraday, known for his scientific genius and entertaining public lectures, was a frequent visitor. On one particular occasion he was invited to come for lunch and "see chloroform administered to Beasts, Birds, Reptiles, and Fishes." These were the early days of experimenting with anesthetics, the first surgery under anesthesia having been performed in 1846, in the United States. Jacko was probably volunteered for the experiments. Being a mammal, he would have succumbed quite rapidly. But birds are fairly resistant to chloroform, and would have taken much longer to become unconscious. It is difficult to imagine how Buckland proposed to chloroform a fish.

The spring of 1848 was marred by fears that the political turmoil that had just erupted on the other side of the Channel might spread to England. The French Revolution was barely fifty years old, but France had again been thrown into bloody conflict. This time the working classes had risen up against the conservative leadership of the Republic. The geological circle received a first-hand account of the turmoil during a dinner party hosted by Sir Robert Peel. The select gathering around the table included Buckland, De la Beche, Lyell, Mantell, and Owen. Lyell wrote to his sister the following day, telling her about the extraordinary happenings:

> Bunsen [the Prussian Ambassador] told us that there are some 30,000 communists in Paris who are for property in common and no marriage. . . . He (Bunsen) blamed [King] Louis Philippe for not having seen . . . that a moderate extension of the suffrage could alone defer some catastrophe. . . . After the ladies retired . . . Sir Robert . . . gave an account of what the French Ambassador had told him just two hours before. . . . .

Mantell had taken his microscope to the party, and shown a few specimens to Lady Peel, but the French revolution had been the main topic of conversation, with a fresh rumor arriving every half hour.

As the revolution in France spread to Italy, the British Chartists planned a mass demonstration in London, to present a petition of their reform demands to the government. Two previous petitions, both demanding male suffrage, had been flatly rejected by Parliament, but the Chartists judged the turbulent times propitious for another attempt. The demonstration would be held on Kennington Common on Monday, April 10, to be followed by a march to the House of Commons to present the petition. A crowd of 150,000 was expected.

The government, fearing the worst, took elaborate precautions to protect life and property in the capital. Public buildings were sandbagged, 7,000 troops were mobilized under the command of the Duke of Wellington, and 150,000 volunteers, from all walks of life, were sworn in as special constables.

The Home Secretary decreed that the Chartists' planned procession to the House of Commons was illegal.

Consternation turned to trepidation in the city as the fateful Monday approached. Queen Victoria left Buckingham Palace for the safety of the Isle of Wight, looking "very ill and pale." Buckland swore in the bigger Westminster boys as special constables, issuing them truncheons. He vowed that if the Chartists attempted to break into the abbey, he would knock down each one, as he entered, with a crowbar. Owen stood on guard at the Royal College of Surgeons. Mantell, who was now living in London, saw cavalry and artillery being quartered at Clapham and at Buckingham Palace.

The dreaded Monday dawned darkly, with heavy clouds and the threat of rain. The city stirred and braced itself, but the streets were as deserted as on a Sunday. There were no soldiers to be seen on the streets because the wily Duke of Wellington had garrisoned them out of sight in houses and gardens.

The crowd that gathered on Kennington Common was far smaller than the Chartists had hoped for. The demonstrators stood around talking among themselves, expectantly waiting for something to happen. Eventually their leader, Fergus O'Conner, stepped forward. As he addressed the party faithful, appealing to them not to harm their cause by any acts of violence, an eagle soared over their heads, flying toward Westminster. This was taken as a good omen.

But the day did not belong to the Chartists. O'Conner capitulated to the authorities and abandoned the planned march on Parliament. Then it started raining so heavily that the speeches were interrupted. The wet and bedraggled crowd broke up, abandoned their soggy banners, and started to make their way home.

Buckland and his children stood at the top of the abbey tower, looking down on the wet London afternoon, though there had been nothing to see so far. But then they caught sight of a cab being driven into the yard of the House of Commons, through the drizzling rain, with the charter tied on top.

The great bird that had been sighted over Kennington Common was Frank Buckland's pet eagle, escaped from its tethered perch in the Deanery garden. The wayward bird flew across the Thames, scanning the city below with its raptorial eye. The day of unrealized expectations was drawing to a close, and the sun, somewhere beyond the clouds, was dipping toward the horizon. The eagle, spotting the familiar sight of the abbey's flying buttresses and towering pinnacles, immediately recognized where it was. Then something of particular interest captured its predatory attention. Wings partially retracted, talons outstretched, the eagle plummeted earthward. As it seized the live chicken Buckland had tethered to a pole for bait, one of its legs was fettered. And so it was that Frank's willful pet was recaptured on the abbey roof.

The charter, like the Chartists' two previous attempts, failed to sway government policy. It would not be until the end of the century that all men won the right to vote. The failure of this third charter marked the beginning of the end of the Chartist movement. Their failure was not because the working class was so weak, but because the middle class—the shopkeepers and businessmen who had prospered from the Industrial Revolution—was so strong.

As 1848 drew to a cold and dreary close a number of Londoners died of cholera, which was nothing unusual. But the disease returned to the capital with a vengeance the following year, reaching epidemic proportions and claiming several hundred lives daily during its height. Buckland, like Mantell, was well aware that the outbreaks were attributable to the lack of proper sanitation and adequate clean water, but his advice to the authorities went unheeded, as did his advocacy of disinfectant.

By November the disease had almost run its course and Buckland conducted a special thanksgiving service in the abbey. He preached a sermon on personal cleanliness, which caused quite a furor. But he saved his harshest words for those he held responsible for the disease—the owners of the disgusting hovels in which the poor were forced to live. He railed against the public officers and city authorities for not providing adequate supplies of clean drinking water, saying it would be the fault of Parliament if the ills were not remedied. With the House of Commons just across the street, it is likely his sermon raised many eyebrows.

Others would have dismissed the tirade as just the rantings of an eccentric. Buckland was certainly that. He had been unconventional all his life, but it was at about this time that his behavior began showing signs of irrationality, even measured against his whimsical yardstick. Initially, incidents in which he beat and scratched himself were isolated and of short duration. His youngest children thought it was all a game and played along with the fun, but the older ones were horrified. His condition steadily deteriorated, forcing him to leave Westminster and retire to Islip. "He puzzles everybody," his wife wrote to De la Beche in the spring of 1850. "There is so much reason & so much non-reason, so much strength & so much feebleness." Writing to De la Beche six months later she told him that her husband was no longer able to feed himself, and was kept alive by "artificial means." But his general health was good—it was his mind that had failed him.

Buckland clung to life, if not to rationality, for six more years, dying on August 14, 1856. An autopsy revealed that the base of his skull and the first three cervical vertebrae were in an advanced state of decay. The erosion of the bone was caused by tuberculosis, which had probably also infected his brain. The inflammation of the vertebrae would have put pressure on his spinal cord, interfering with the normal flow of cerebrospinal fluid, causing pressure changes in his brain. This alone would have caused the symptoms of mental illness. However, it is likely that his brain, too, would have been directly affected by the infection. According to the autopsy the brain appeared normal, but it is quite possible there were tubercular nodules that were overlooked. These would have added to his neurological symptoms.

The pathological portions of Buckland's mortal remains are on public display in the galleries of the Hunterian Museum, London. It would have pleased him to know he would continue teaching students long after his death.

Buckland's legacy to paleontology is *Megalosaurus,* the first dinosaur ever named. But this major historical figure will probably always be remembered as the charismatic eccentric who liked to play to the crowd.

Conybeare followed his old friend to the grave a year later, aged seventy. Buckland's wise counsel and academic mentor, Conybeare's lasting paleontological contribution is the establishment of the plesiosaurs as a distinct group of reptiles and his naming of many plesiosaurian and ichthyosaurian species.

De la Beche predeceased Buckland by a year, at the age of only fifty-nine. His major contributions were to geology rather than paleontology, spearheading the Geological Survey of Britain. But perhaps his most enduring legacy is the satirical cartoons and sketches he penned of his geological colleagues, and the times in which they lived.

Mantell had died in 1852, four years before Buckland. He slipped into unconsciousness after taking an extra dose of opiates to alleviate his back pains, and was buried beside his beloved daughter, Hannah. At his own request no invitations were sent out to attend his funeral. The small gathering beside his grave included his older daughter, but not his estranged wife. As Robert Bakewell had prophesied, Mantell is immortalized by *Iguanodon,* the second dinosaur ever named. But for me he will be remembered as the driven man, so obsessed by his unending quest for the unobtainable that he lost the love of a beautiful wife, and almost lost his children too.

Richard Owen, with one of his grandchildren, in his twilight years.

Hawkins died in 1889, at the age of seventy-nine. He is buried on the Isle of Wight, where he spent the latter half of his life. During his early days on the island, he married a woman fifteen years his senior, but she appears to have moved back to London soon after their wedding, and died three years later. Even more odd, he adopted an adolescent boy when he was in his early fifties. This geological gadfly, who caused so much trouble wherever he went, obviously suffered from a serious mental disorder. This is apparent from his irrational behavior and from the manic ramblings of his writings. His everlasting contribution to paleontology is the magnificent ichthyosaurs and plesiosaurs that still grace the public gallery of London's Natural History Museum. But their seeming perfection will always raise questions regarding their authenticity.

Lyell died in 1875, after a short illness, at the age of seventy-seven. His *Principles,* a cornerstone of geology, influenced generations of students, including Darwin. Aside from using the book as a guide to interpreting geological phenomena, Darwin used Lyell's principle of interpreting the past by reference to the present to arrive at his theory for the origin of species. Lyell's principle still has currency today.

Darwin, for all his concerns about ill health, saw his seventy-third year, dying in the spring of 1882. His stellar contribution was to provide a plausible and testable mechanism—natural selection—to account for the appearance of new species on Earth.

Owen outlived them all. He died in 1892, at the ripe old age of eighty-eight. But he was a lonely old man. His wife had died almost twenty years before, and their only son had committed suicide, by drowning himself in the river Thames, at the age of forty-nine. But Owen still had his seven grandchildren, and he lived with them, and his daughter-in-law, at Sheen Lodge, on the outskirts of London. This magnificent home, with its spacious grounds, was granted to him by Queen Victoria in 1852.

London's Natural History Museum, which he helped build, was opened in 1881. This was the realization of his dream of a national collection of natural history. He retired two years afterward, and was honored the following year by a knighthood and a grand dinner at the museum. But he does not appear to have had many visitors at Sheen Lodge during the last few years of his life. Many of his friends and associates had already died—one of the disadvantages of longevity—and the younger biologists and paleontologists seemed to have had little time for him. This has been attributed to his rejection of transmutation, and his bitter feuds with Darwin and his supporters. But he left a monument of scientific publications. And, of course, he left us the concept, and name, *dinosaur.*

Mary Anning, the unsung heroine of Lyme, had been the first loss among the fossilists. She died of breast cancer on March 9, 1847. She was buried in the Parish Church of Lyme Regis, within sight and sound of the sea. The year following her death, her lifelong friend, Henry De la Beche, concluded his anniversary address as president of the Geological Society with a short eulogy, later published in the *Quarterly Journal of the Geological Society.* He referred to her as one who had not been "placed among even the easier classes of society, but who had to earn her daily bread by her labour. . . . " He acknowledged how she "contributed by her talents and untiring researches . . ." to knowledge of marine reptiles, and other animals, and how her talents and good conduct had won her many friends. The obituary was quite short, but it was remarkable in itself because she was not a member of the society—the society did not admit women members until 1904. She was commemorated by her hometown two years later with the installation of a stained-glass window in the Parish Church.

16

Beside the Sea

Two hundred years after Mary Anning's birth an international delegation of geologists, paleontologists, historians, fossil collectors, authors, and other interested parties gathered in Lyme Regis to celebrate her life and times. For four days we met in the Marine Theatre, reviewing a broad tapestry of topics, ranging from Mary Anning's paleontological contributions to the unstable geology of the Dorset coast. The small theater overlooks the sea and was built on the site of the old public baths, which she would have passed on her way to the beach. There was some doubt whether the small building would accommodate so many participants, but we managed, cozily, saturating the local tea rooms during coffee breaks. The meeting concluded, most appropriately, with a strawberries-and-cream garden party, hosted by Lyme's celebrated novelist, John Fowles.

If the essence of the meeting were distilled into a single sentence it would be that Mary Anning happened to be the right person, in the right place, at the right time. If she had not had an eye for fossils or the tenacity to keep on searching even when it seemed hopeless, she would not have made so many outstanding discoveries. Had she lived miles inland, or miles away from this particular stretch of coastline, there would have been no fossil reptiles for her to discover. If she had lived a few decades earlier the significance of her discoveries would have been lost on the learned gentlemen of the time. And if she had been born a few decades later, someone else probably would have made the discoveries before her.

Being a woman robbed her of much of the kudos she would have enjoyed had she been born a man. She resented this, as she recounted to Anna Pinney, a young woman she took on several fossiling forays: "She says the world has used her ill . . . these men of learning have sucked her brains, and made a great deal by publishing works, of which she furnished the contents, while she derived none of the advantages."

Nonetheless, she did enjoy celebrity for much of her adult life, if not the heroic acclaim attaching to her name today. And what might she have said if she had witnessed the 1999 meeting held in her honor? "At last," John Fowles offered.

Mary Anning lives on. Not only in memory, and in the legacy of the fossils she discovered, but also in the work of the contemporary collectors of Lyme. Their participation in the meeting forged a living link with her distant past. The exhibition of their most recent finds gave a hint of the promise for the future.

David Costin lives off Anning Road, in Lyme Regis, which seems very apt for a man who has spent most of his life collecting and preparing fossils. He does not do as much collecting now as he once did. Instead, he spends most of his time preparing fossils—mostly marine reptiles—in the small workshop at the bottom of his garden.

March 2, 1982, is a date he will never forget. Not because of any spectacular fossil find, but because of a serious car accident that put him into the hospital with a badly fractured leg. The leg took an age to heal, and as the weeks dragged on he became increasingly anxious to get back into the field. His collecting companion, Peter Langham, visited from time to time, and on one of these occasions David persuaded Peter to take him for a trip. His leg was still not completely healed, but he desperately needed to start collecting again.

They decided to take an hour's drive north, to prospect for fossils along the Somerset coast, on the Bristol Channel. The same Blue Lias outcrops there as at Lyme Regis, but, because of its more remote location, it attracts fewer collectors. The chances of finding fossils is thereby increased. Acres of liassic reefs are exposed at low tide, but much of this is often covered by marine algae and by the silt carried down by rivers. But occasionally this covering is scoured off by the currents, which was the case on this occasion.

As they walked along the desolate shoreline, David noticed a series of depressions in the shale. These turned out to be impressions left by a series of vertebrae and ribs. The elements themselves had long since weathered away, but when they took a closer look they found bone there too. Peter was not very optimistic about the specimen because the bone was so poorly preserved. It looked like rotten wood and there seemed little point in trying to collect the specimen. But David was determined: He needed something to keep him occupied.

Within a few hours they had collected most of what remained of the skeleton. This comprised a skull, one complete forefin, most of the shoulder girdle, and a short series of vertebrae and ribs. They returned a few days later and collected a string of articulated vertebrae and ribs.

The bone was so badly weathered on the exposed side that David decided to flip the slab over and prepare the skeleton from the other side. Progress was painfully slow because the bone was so fragile. Acid could not be used and the matrix had to be scraped away a little at a time using small hand tools.

As more of the skull was uncovered it became apparent that it had a remarkably long and slender snout. It might have been the long-snouted species *Leptonectes tenuirostris.* But as preparation proceeded, David noted something unusual: The snout continued well beyond the tip of the lower jaw. A small overbite is not unusual in *L. tenuirostris,* but this long snout just kept on going. Realizing he had found something new, he notified Peter Crowther, who was then the curator at the City of Bristol Museum and Art Gallery.

The original Bristol museum once housed a large collection of marine reptiles, mostly collected during Anning's time. Unfortunately, most of these were destroyed during World War II, and Crowther had spared no effort to build up the Bristol collection again. To this end he made arrangements to acquire the new ichthyosaur. Once this had been achieved, I was invited to study it.

As soon as I saw the specimen I realized it was so different from all other kinds that a new generic name had to be established for its reception. As mentioned earlier, this new ichthyosaur, named *Excalibosaurus,* appears to bridge a gap between long-snouted ichthyosaurs like *L. tenuirostris* and the swordfishlike *Eurhinosaurus.*

Scientific names have two parts. The first part is the genus, or generic name, and starts with a capital letter; the second is sometimes called the trivial name. The entire combination is referred to as the species, or specific name. As lamented by John Fowles in *The French Lieutenant's Woman,* Mary Anning never had a new species named after her until long after her death. The failure to recognize her remarkable saurian discoveries in this way was an unfortunate oversight. This was not to be the case for the new ichthyosaur, which I would name after its collector.

When a personal name is used for a species it has to be Latinized, by giving it an appropriate ending, according to established grammatical rules. Needless to say, care has to taken to ensure the correct spelling of the original name. I first met David during my visit to England to study the specimen. Aside from discussing the find, I wanted to check the precise spelling of his surname. My understanding was that his name was David Costain, because that is the name by which he was known in Lyme. To my surprise he said his name was actually Cost*in,* not Cost*ain.* When I asked why everyone called him *Costain,* he explained that when his new wife signed the church registry on their wedding day, she inadvertently wrote her new name with an "a" before the "i"; he thought it best just to leave things that way. He assured me his name was spelled COSTIN on his birth certificate. And so it was that the new discovery was officially named *Excalibosaurus costini,* in a paper published in 1986.

A year or so after the new name appeared in print, I was again in England, doing the rounds of the local collectors in Lyme, and visited my friend John Fowles. John, with his passionate interest in the geology and paleontology of Dorset, had read the *Excalibosaurus* paper I had sent him. As we talked about the species, he casually asked why I had named it *Excalibosaurus costini* instead of *Excalibosaurus costaini.* When I explained the nuances of David's real name, I thought he sounded less than completely satisfied. That worried me because John has lived much of his life in Lyme, and has known the local collectors for years. Perhaps I had made some ghastly mistake. I checked the local phone book and found the name Costain, which did not allay my mounting anxiety. I then called on David, hoping to verify my recollection of our previous conversation about his name. He reassured me that his name really was spelled Costin, in spite of the fact that everybody in Lyme, including the telephone company, knew him as Costain. All this fretting over a name may strike some as esoteric excess, but once a name has been published it is a permanent part of the scientific literature, so I was greatly relieved that the spelling was correct.

David Sole, another local collector, gave a presentation at the Anning meeting on recent finds along the Dorset coast. His primary interest is ammonites, and he owns a stunning collection of these spiral-shelled fossils. But he has also made some remarkable ichthyosaur discoveries, often while looking for ammonites. During the winter of 1987 he was prospecting the coastal cliffs near Charmouth. His attention was focused on a particular band of shale in the cliffs, known locally as the "Topstone Bed" and noted for its nodules. The reason for his interest was that the nodules, which are usually only an inch or two thick, sometimes contain ammonites. As he peered up at the shale band, some seventy feet above the beach, something unusual caught his eye: a remarkably thick nodule. He had to have a closer look.

By cutting a series of footholds in the cliff face he managed to climb up to the level of the nodule. When he loosened it from the cliff and took a careful look, he discovered it contained bone. He immediately started digging into the shale and found more of the same kind of nodules. He had discovered the remains of a large ichthyosaur, but at this stage he did not know how much was there.

He returned the following day with David Costin and another collector, and they began to dig into the cliff. The more they dug the more they found—not only nodules containing bone, but also bones that were not encased in limestone. And the specimen just kept on going, farther and farther and deeper and deeper. So they kept on digging farther and deeper into the cliff, creating a substantial excavation.

Digging a large excavation in such a soft and crumbling shale can be a hazardous business. It is even more risky when the excavation is being made seventy feet above the ground. Without warning a mass of material at the back of the excavation slumped forward, partially burying Costin and forcing Sole to rush headlong down the cliff, closely followed by falling debris. Their other friend scrambled clear just in time. Fortunately nobody was hurt, and an important lesson was learned.

The only safe way to reach the specimen was to dig down from above, sloping the sides and back of the excavation for greater stability. The land along this section of the coast belongs to the National Trust, a privately owned conservation authority, and Sole first had to obtain their permission to excavate the cliff. Then, working almost entirely alone, he began the mammoth task of digging a large quarry down to the level of the specimen. What made the undertaking all the more daunting was that he did not know the extent of the find—it could turn out to be nothing more than a string of vertebrae and ribs. It took him from November until February to complete the task—the coldest and wettest months of the year. When, at last, he reached the level of the specimen,

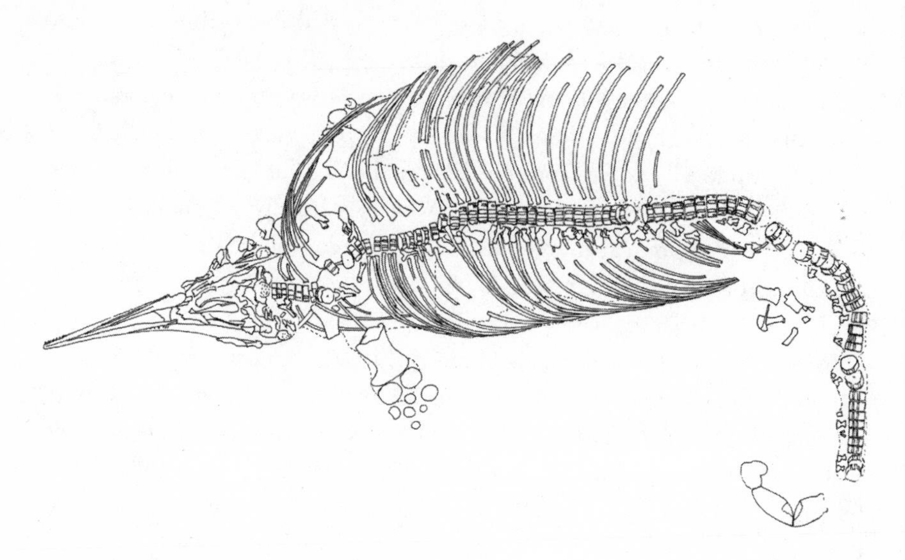

The skeleton of *Leptonectes solei,* named in honor of its discoverer, David Sole.

he enlisted the help of David Costin and Chris Moore, another experienced collector, to remove it.

Fortunately, the specimen turned out to be an almost complete skeleton. Most of the bones were embedded in limestone nodules, but some, including most of the ribs, were not so encased. Before removing any of the parts they made a sketch of their disposition, to facilitate the reassembly of the skeleton after its preparation. There was no access to the top of the cliff, so the larger parts of the skeleton had to be lowered by rope to the beach below, and loaded onto a Landrover.

The specimen was delivered to Costin's workshop, where he worked for several months removing the overlying matrix from the bone, using an Airscribe, and dilute acetic acid. The skeleton, though lacking much of its tail, was over twenty feet long and they had to rent a local hall so they could spread it all out. Crowther went to see the specimen and was so impressed that he began a fund-raising campaign to purchase the skeleton for the Bristol museum. It was at about this time that the discovery attracted the attention of the media, inundating Sole with newspaper and television interviews.

My studies of the slender-snouted giant confirmed that it was a new species, which I named *Leptonectes solei,* in honor of its collector. Like *Leptonectes*

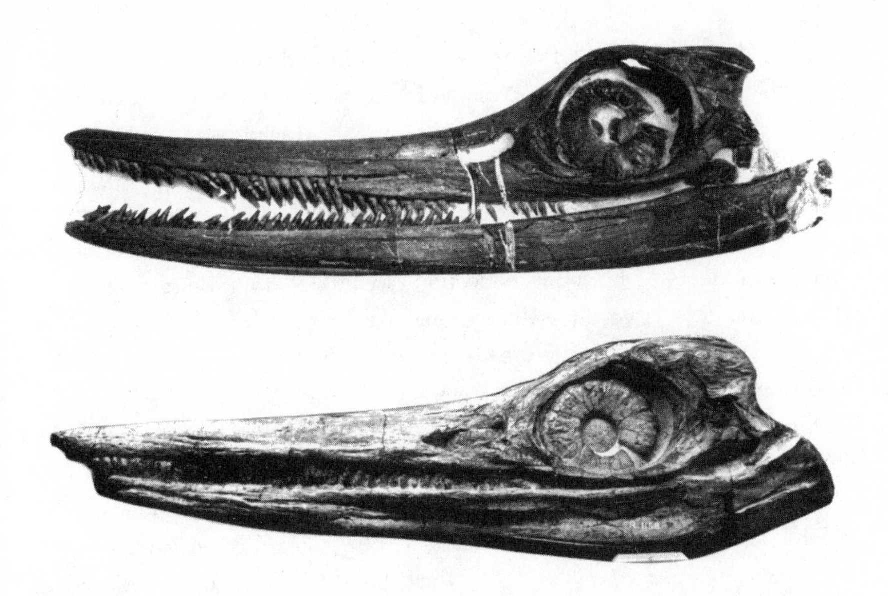

The three-dimensional skull of *Temnodontosaurus platyodon* (top) that David Sole found at Black Ven. (Skull length 32 inches; 81 centimeters). Mary Anning's first ichthyosaur was more than 50 percent larger.

tenuirostris, it has a remarkably long and slender snout, but the skeleton is considerably larger. The largest specimen of *L. tenuirostris* has a total length of a little more than twelve feet (3.8 meters), whereas this skeleton is about twice that length. It is unfortunate that the specimen is not complete, and that the skull and fins are rather poorly preserved because few comparisons can be made with *L. tenuirostris*.

A year after this impressive discovery at Charmouth, Sole made another important find. Visiting Lyme with his brother-in-law, they walked along the beach to Black Ven, where Anning did much of her collecting. Sole was interested in collecting some of the nodules—known locally as "flat stones"—that lie embedded in the marl because they sometimes contain exquisitely preserved ammonites. Using a thin steel rod to probe the soft shale, his brother-in-law located a small nodule. When they took a closer look they found it contained an ichthyosaur vertebra. They dug into the loose marl hoping to find more ichthyosaur bones, but were disappointed.

Sole kept checking the area for the next few weeks, but without success. He then tried digging a trench in the general vicinity of the find, and was rewarded with another vertebra. Spurred on by this find he did some more digging, and began to unearth a succession of articulated vertebrae. These seemed to be leading toward the front end of the animal, which was headed into the cliff. With mounting anticipation he dug deeper into the marls of Black Ven.

He told me afterward that following the skeleton into the cliff, knowing that every shovelful of marl was inching him closer to the skull, was the most exhilarating experience of his entire collecting career. He was not disappointed in what he found: a large three-dimensional skull embedded in a nodule. The limestone layer was only a few inches thick, so he could see its entire outline—it was complete, right to the very tip of its snout.

The new skeleton did not prove to be a new species. But it was remarkably well preserved, and resolved a long-standing puzzle relating to the species to which Anning's first ichthyosaur belonged. Anning's four-foot-long skull, which she collected when she was twelve, belongs to the species *Temnodontosaurus platyodon*. This killer-whale-sized ichthyosaur reached a total body length of about thirty feet (9 meters).

Some years ago I described a similar but somewhat smaller species, which I named *Temnodontosaurus risor.* The second part of the name, *risor,* meaning "mocker" in Latin, was chosen because the slightly upturned curvature of its rostrum imparted a sardonic expression to the skull. Sole's new specimen had a similarly curved rostrum, identifying it as *T. risor.* What made this particular skull so unusual was that the jaws gaped open, accentuating the jocund appearance of the upturned snout. Another striking feature was that the skull, which is a little less than three feet (81 centimeters) long, seemed remarkably large relative to the rest of the skeleton. Large heads, as mentioned earlier, are characteristic of immaturity, making me suspect this was a juvenile individual. Could *T. risor* merely represent juvenile individuals of *T. platyodon*? To check this I arranged the data for the other specimens of *T. risor,* in order of increasing size, along with size-ordered data for *T. platyodon.* It was quite apparent from the regular trends in the data that the individuals formed part of a continuous growth series. For example, the smallest individuals had the largest orbits, relative to skull length, the same way that kittens have relatively large eyes. But as body size increased, the orbit became relatively smaller. *T. risor* was clearly a young *T. platyodon,* and the name was accordingly relinquished.

Peter Langham first became interested in fossil collecting through his father, Bob. The family lived in Reading, some forty miles from London, and they used to travel down to Lyme Regis during the summer to spend fossil-collecting holidays. Many years later Peter took up permanent residence in Lyme, and his father continued paying visits. And it was during one of their collecting trips together that they made one of their most important discoveries.

Father and son, accompanied by David Costin, had walked along the coast to Pinhay Bay to collect a small ichthyosaur the Langhams had found a few days earlier. They were making their way back toward the town, carrying the specimen between them, when they decided to stop for a rest and a smoke. When Bob had finished his cigarette he idly tossed it onto the liassic beach, now laying beneath several inches of water. Casually glancing down to see where it had landed, he caught sight of what appeared to be a series of ribs, buried in the shale. Running his shovel along them produced a rhythmic tinkling—they really were ribs, and belonged to a fairly large ichthyosaur.

The thin band of shale in which the specimen was embedded lay beneath a limestone ledge. This had to be removed before they could start scraping away at the shale to see how much of the beast they had found. Working in water added to their difficulty, but they managed to remove the limestone slab, collecting most of the underlying specimen that same day. Returning three more times at low tide, they recovered the rest of the skeleton, save for the very tip of the snout, which they could not remove from the rock. It may still be there, embedded in the shale, if the sea has not already destroyed it.

The specimen turned out to be an almost complete skeleton, over twelve feet (3.8 meters) long. The snout was long and thin, and the body was strongly arched, as if the animal had been captured in freeze-frame jumping out of the water. From that time on it became known as the "leaping ichthyosaur." I first saw the specimen when it was on temporary display at the Philpot Museum in Lyme Regis. At that time John Fowles was the honorary curator of the public museum. The specimen was later moved to "Dinosaurland," a private museum in Lyme, founded by Peter Langham and his wife Cindy. When they sold the museum, several years later, they put the "leaping ichthyosaur" on the market. The specimen was not a new species, which is probably why little interest was shown by any of the museums in Britain. However, it is of considerable scientific value because it is the largest exemplar of *Leptonectes tenuirostris.* As there were no domestic offers to purchase the specimen, I felt free to try and obtain it for my museum. After several unsuccessful attempts to obtain funding, I was eventually able to purchase it through a generous bequest made to the Royal

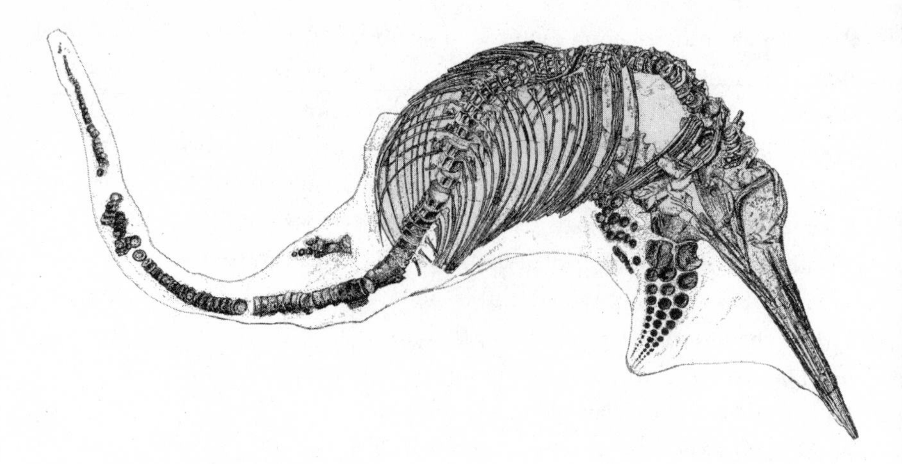

The "leaping ichthyosaur," discovered by Peter Langham's father, Bob Langham.

Ontario Museum by one of its patrons, the late Louise Stone. The specimen arrived in 1999 and now resides permanently in Toronto.

Although *L. tenuirostris* is one of the most common ichthyosaurs, most of the specimens are fragmentary, comprising partial forefins, isolated humeri, and rostral segments, with precious few skeletons. The most perfect skeleton, the type specimen, is from Street and has artificial matrix, raising serious doubts about its authenticity. The "leaping ichthyosaur" has a minor amount of plaster restoration, but Peter Langham pointed out the extent of this from the outset. So, aside from being the largest exemplar of the species, it is probably also the most complete.

A ten-minute drive from Charmouth, down a narrow winding lane, brings a visitor to Seatown. The name is rather a misnomer because there is little there beside a few houses, a car park, and the *Anchor* pub. The shingle beach does not have much appeal for families with small children, but the hills, readily accessible from the beach by a footpath, offer the most magnificent views of the surrounding countryside. And those interested in fossils cannot fail to find something worthwhile embedded in the marl that outcrops toward the top of the beach. The most common fossils are the belemnites, the long bullet-shaped shells of animals related to squids. Indeed, the marl is so peppered with their remains that it is called the Belemnite Marls. The Belemnite Marls are geolog-

The incomplete skeleton of *Leptonectes moorei,* named in honor of its discoverer, Chris Moore.

ically younger than the Blue Lias by several million years. Until the beginning of 1995 ichthyosaurs had never been found there.

The Dorset coast, no stranger to winter gales, was lashed by a particularly heavy storm one January night in 1995. Chris Moore, who owns a fossil preparation workshop in Charmouth, listened to the howling wind, hoping the sea would uncover some fresh fossil-bearing exposures. He was not disappointed. When he visited Seatown the following day he found the storm had scoured most of the shingle from the beach and from the foot of the cliffs. As he walked over the freshly exposed acres of marl, he noticed a small string of vertebrae and ribs. He had discovered the front half of a remarkable new ichthyosaur.

I first saw the new specimen in David Costin's workshop, soon after it had been prepared. It is a small ichthyosaur—the skull is just over one foot (33 centimeters) long—and although not fully mature, it appears to belong to a small species. I was struck by its remarkable resemblance to *Leptonectes tenuirostris.* Like that species, its forefin has only four digits, the individual bones of which are relatively large, and their humeri are identical. The skull has a relatively huge orbit, with a slender snout. However, in stark contrast to *L. tenuirostris,* the snout is very short.

Aside from its scientific interest, this exquisitely well-preserved specimen is aesthetically pleasing. My colleague Angela Milner, of London's Natural History Museum, acquired the specimen, and we agreed to publish a joint paper on its description. We named the new species *Leptonectes moorei,* in honor of its collector.

The discovery of *L. moorei* adds an entirely different piece to the picture that was beginning to emerge. Prior to the discovery it was thought that the *Leptonectes* group of ichthyosaurs, which includes *Excalibosaurus* and *Eurhinosaurus* as well as *Leptonectes,* were all specialized toward rostral elongation, where the snout is greatly extended. However, we now know that there were two contrasting patterns of rostral evolution: one leading to extreme elongation, as exemplified by *Eurhinosaurus* with its swordlike bill, and the other to the much abbreviated snout of *L. moorei.*

The modern collectors of Lyme, like their celebrated predecessor, are exemplary of what professional fossil collecting can and should be. They spend long, tiring hours searching the rocks on a regular basis, in fair weather and foul, throughout the entire year. Museum curators and university researchers, in contrast, can only spend a fraction of their time in the field. Fossils are continually weathering out of the rocks, and if they are not collected they are soon lost and destroyed. It follows that if fossil collecting were the sole preserve of museums and universities, most of the fossils from the Dorset coast would be lost. The local collectors are the best guarantee against this attrition, and it is no coincidence that many of the most important fossils that have ever been found were collected by locals.

Ever since Anning's time professional collectors have borne something of a stigma for the commercialism they bring to the esoteric domain of fossils. In recent times they have been held partly or wholly responsible for the astronomic rise in the prices of fossils. It is true that some collectors do treat fossils as mere commodities, to be sold to the highest bidder. And some commercial dealers are quite unscrupulous in flaunting collecting restrictions and export regulations. One of the most notable examples of this pertains to China's rich fossil heritage. Chinese fossils, including dinosaur eggs and Cretaceous birds, have become familiar sights at international fossil shows, but, in spite of assurances to the contrary by the dealers, almost every one has been illegally exported from China.

Although there are no export restrictions on fossils in Britain, most of the Dorset collectors make every effort to sell significant fossils—those that may be new species or that are rare—to British museums. In recent years a pilot project was launched for the West Dorset coast, to promote responsible and safe collecting. Included in the code of conduct is the registration of important specimens, with full locality and geological data, so that paleontologists can be kept informed of what has been discovered. Collectors interested in selling such fossils are first to offer them to museums before approaching the private sector. But this is something many of the Dorset collectors have been doing for years.

If the Dorset collectors placed their vertebrate fossils on the open market they would obtain much higher prices for them. Instead, they try to make them affordable to public museums. There is good reason for doing this, aside from the philanthropy of making their fossils available to everyone. This has to do with the *International Code of Zoological Nomenclature*. This code is the system of rules and recommendations that researchers must follow when erecting new species. According to the code, the reference specimens on which new species are based—referred to as *type specimens*—must be deposited in a public museum or similar institution. This is to ensure both their safekeeping and their accessibility to researchers. If specimens were kept in private hands they could not be used to erect new species, nor could they be used for comparison with those already named—they would be essentially lost to science.

Seldom does a year go by without my receiving a phone call or letter from one of the collectors of Lyme, telling me of their latest find. At the time of writing I have just returned from England, where I examined Chris Moore's latest discovery: another small ichthyosaur from the Belemnite Marls that appears to be something entirely new.

One day, as I frequently remind the local collectors, I expect to hear they have unearthed the "big one." I know this unidentified behemoth lies there somewhere, hidden in the shales, because I have found tangible clues of its existence. The story of the discovery of this elusive giant began in 1995 when I was visiting the Philadelphia Academy of Natural Sciences. This was the second time I had looked through their ichthyosaur collection, having been there some ten years earlier when the academy hosted the annual meeting of the Society of Vertebrate Paleontology (SVP). The academy's ichthyosaur collection, which is not extensive, was acquired during the nineteenth century and contains some historically interesting specimens. Although they do have some complete skeletons, most of the material comprises isolated bones, neatly housed in wooden cabinets. As I pulled open one drawer after another I came upon a large bone, about the size of a dinner plate and shaped something like a cookie with a large bite removed. One end of the bone was considerably thicker than the other, and swollen into a large knob that my hand could barely span. The attached label identified the specimen, from Lyme Regis, as an ichthyosaur coracoid bone. The coracoid is part of the shoulder girdle, and one of the larger bones in the ichthyosaurian skeleton. Given that the bone was so large, and the same shape as a coracoid, made the identification seem entirely reasonable—it certainly seemed reasonable to me ten years earlier. But as soon as I saw it for the second time, I knew its true identity. The robust element was not part of the shoulder at all but part of the skull, namely a quadrate bone. The quadrate is the bone at the back of the skull to which the lower jaw attached, and the large knob was its articular surface. The quadrate is a far smaller bone, relative to the length of the skeleton, than a coracoid, and this was the largest ichthyosaurian quadrate bone I had ever seen. The entire animal must have been colossal—larger than any ichthyosaur ever found in England. To get an idea of its size, I made comparisons among some skulls and their quadrates. There is a huge isolated ichthyosaur skull of *Temnodontosaurus platyodon* in London's Natural History Museum that is just over six feet (1.84 meters) long. Such a large animal would be about thirty feet (8.3 meters) in length, the size of a modern killer whale. Its quadrate is eight inches (20 centimeters) high, compared with eleven inches (27 centimeters) for the Philadelphia quadrate. By simple scaling this would give an estimated body length for the unknown ichthyosaur of forty feet (11.2 meters), which is the size of a modern rorqual, or baleen, whale.

Alerted to the existence of a very large, but unidentified ichthyosaur in the English Lias, I visited the Natural History Museum in London and checked through their extensive collection. After some hours of searching I discovered a massive scapula (shoulder blade) that was eighteen inches (44.4 centimeters) long, and an isolated vertebral centrum eight inches (20.5 centimeters) in diameter. Simple scaling from much smaller skeletons of *T. platyodon,* the largest English ichthyosaur, gave estimated body lengths of forty-six feet (14 meters) and forty-nine feet (15 meters), respectively. I also found some massive isolated teeth, quite unlike those of *T. platyodon*. The largest teeth I have seen of this species are just over two inches (5.4 centimeters) long, whereas the isolated teeth were nearly five inches (11.6 centimeters) long. Unlike the teeth of *T. platyodon,* where root and crown are of equal length, the crowns of the isolated teeth comprise only about one-third of the total length of the tooth. Furthermore, they had sharp cutting edges. Similar cutting edges can be seen in some *T. platyodon* teeth, but their crowns are relatively much longer, and the teeth are, of course, considerably smaller.

I suspect the massive isolated teeth belong to the unknown giant. This ichthyosaur was as large as *Shonisaurus,* the largest known ichthyosaur, from the latest Triassic of North America. All we know so far about the enigmatic giant is that it reached lengths upward of about fifty feet (15 meters), and had massive teeth, with sharp cutting edges and short crowns. When a skeleton is eventually discovered it will be the largest ichthyosaur ever found in England.

New species of fossils continue to be discovered at a rate surpassing that of the early days of paleontology. The major groups of animals have already been discovered, and the chances of a major find, equivalent to Anning's discovery of the plesiosaurs, is remote. Nonetheless, there is no dearth of surprises. Dinosaur discoveries have tended to capture the headlines, like the recent discoveries of small feathered dinosaurs from China. But there have been some remarkable nondinosaurian finds too. For example, there have been some fantastic pterosaur finds, like *Quetzalcoatlus,* which was unearthed in Texas. With an estimated wingspan of up to forty feet (12 meters)—the size of a twin-engine light plane—this was the largest animal ever known to fly. *Pterodaustro,* a heron-sized pterosaur discovered in Argentina, is remarkable for its long, curving bill that was filled with bristlelike teeth for sieving food items. Some of the new pterosaur finds from South America, like those from the Santana Formation of Brazil, are in remarkably good condition. Not only has the bone been exquisitely preserved, but so too has the microscopic structure of the skin. Even more remarkable was the discovery of some Jurassic fish fossils in which microscopic details of the muscle cells had been preserved.

Many small dinosaurs have been discovered in China in recent years that bear an array of external structures. These range from the furlike "fuzz" of *Sinosauropteryx* to the indisputable feathers—with rachis and barbs—of *Caudipteryx.* I have recently had the opportunity to visit several of China's exciting new fossil localities, and have seen some of the remarkable dinosaurs and birds they have yielded. This has consolidated my conviction of the close relationship between birds and dinosaurs, and the difficulty of trying to draw a line between the two.

Much of the excavation at these localities is being carried out by paleontologists from Chinese institutions, including the Institute for Vertebrate Paleontology and Paleoanthropology (IVPP) and the Geological Museum of

China. But most of the collecting is being done by local farmers. Like the quarrymen of nineteenth-century Lyme Regis, they have found that excavating fossils can be more rewarding than their regular job. Like Thomas Hawkins, a rare few have taken to improving on nature by adding missing parts from other specimens. One such composite, of the birdlike dinosaur *Archaeoraptor,* caused quite a sensation when it was discovered that the tail belonged to another animal. Some opponents of the bird-dinosaur thesis were delighted to see such a dinosaur shot down in flames. Others made much of the fact that professional paleontologists had been fooled. But I think both parties were missing the point. One composite birdlike dinosaur does not make them all suspect, any more than one faked Rembrandt renders all paintings by that artist forgeries. The fact that professionals were taken in by the fossil is unfortunate, but not incomprehensible. People have been fixing up fossils for over 200 years and some are very skilled in the art. We try to be vigilant, and most of the time we are able to spot the compromised specimens, but we are not infallible.

With all the hundreds of species that have been unearthed from the Age of Reptiles we could get the impression that most of what has to be discovered has already been found. In reality, we have only scratched the surface, and the future for paleontological discoveries looks very bright. We can therefore look forward to exciting times ahead. Many of the most important finds will be made by those who are not employed as paleontologists—the spirit of Mary Anning lives on.

Notes

Chapter 1

p. 2 Darwin's quote on his orthodoxy is from:

Darwin, F., ed. *The life and letters of Charles Darwin, including an autobiographical chapter.* London: Murray, 1887, pp.307–308.

pp. 7–8 Darwin's quotes regarding the advantages of delaying publication are from the previously cited reference, pp. 87–88. The citation for the *Origin* is:

Darwin, C. *On the origin of species by means of natural selection, or the preservation of favoured races in the struggle for life.* 3d. ed. London: Murray, 1861.

p. 10 Quotation about zeal from:

Woodward, H. B. *The history of the Geological Society of London*. London: Geological Society, 1907, pp. 15–16.

Chapter 2

p. 16 Quotation from:

Murray, J. "The late Miss Mary Anning." *Mining Journal* 17 (December 25, 1847).

p. 19 The Silvester quotation is from:

Welch, E. "Lady Silvester's tour." *Devon and Cornwall Notes and Queries* 32, pt. 9 (1976): 262–264.

The quotation about cuttlefish was contained in a letter from Mary Anning to William Buckland, December 21, 1830, cited in:

p. 20 Edmonds, J. M. "The fossil collection of the Misses Philpot of Lyme Regis." *Proceedings of the Dorset Natural History and Archaeological Society* 98 (1978): 43–48.

The quotation from the diarist is contained in an extract from the journal of Thomas Allan for June 25, 1824, cited in:

Lang, W. D. "Mary Anning (1799–1847) and the pioneer geologists of Lyme." *Proceedings of the Dorset Natural History and Archaeological Society* 60 (1939): 142–164.

pp. 20–21 Letter from Mary Anning to William Buckland, June 1832: Buckland Papers, Oxford University Museum of Natural History.

The second quotation is from:

Gordon, E. O. *The life and correspondence of William Buckland*. London: Murray, 1894, pp. 7–8.

p. 23 Home's view that it was not "wholly a fish . . ." is from:

Home, E. "Some account of the fossil remains of an animal more nearly allied to fishes than any other classes of animals." *Philosophical Transactions of the Royal Society of London* 104 (1814): 571–576.

p. 26 The quotation about Buckland's geniality was written by Lord Playfair and cited in Gordon 1894, p. 172.

The quotation about Buckland's horse is given in the same reference, pp. 21–22.

p. 27 Although Buckland may have dined at the Three Cups, he used to stay in rented rooms in the town.

Chapter 3

p. 29 The first quotation is attributed to R. Murchison, cited in Gordon, 1894, p. 10.

The second quotation is attributed to H. Acland, cited in Gordon, 1894, p. 31.

p. 30 The comment about Buckland gone to Italy is cited in Gordon, 1894, p. 10.

Shelley quotation from:

Hogg, T. J. "The life of P. B. Shelley." Cited in Edmonds, J. M., 1991. *Vindiciae Geologicae,* published 1820; "The inaugural lecture of William Buckland." *Archives of Natural History* 18 (2): 255–268.

p. 33 William Buckland to the Reverend J. Buckland, October 26, 1818. From a handwritten transcript. Buckland Papers, Oxford University Museum of Natural History.

William Buckland to the Prince Regent, undated typewritten transcript: Buckland Papers, Oxford University Museum of Natural History.

The quotation about Buckland's nervousness was cited in Gordon, 1894, pp. 27–28.

p. 34 The two quotations from Conybeare's letter to Buckland are given by Edmonds, 1991, pp. 259 and 261, respectively.

pp. 36–37 Quotations from:

Buckland, W. *Vindiciae Geologicae; or the connexion of geology with religion explained.* Oxford: Oxford University Press, 1820.

Buckland's reference to vindicating geology is from a typewritten transcript of a letter he wrote to Lady Cole, May 18, 1820. Archives of the National Museum of Wales, Cardiff (84.20G.D153).

p. 40 Charles Laurillard to Georges Louis Duvernay, September 30, 1817, quoted in:

Taquet, P. "Cuvier—Buckland—Mantell et les Dinosaures." In *Actes du symposium paléontologique G. Cuvier, Montbeliard,* ed. E. Buffetaut, J. M. Mazin, and E. Salmon, trans. S. Andrews, pp. 475–491.

Chapter 4

p. 42 The quotation about long journeys is from:

Mantell, G. A. *Petrifactions and their teachings.* London: H. G. Bohn, 1851, p. 207.

His reference to Cuckfield fossils is from:

Mantell, G. A. Unpublished journal; typed transcript, in the library of the Sussex Archaeological and Historical Society, Lewes, in 4 vols. (original in Turnbull Library, Wellington, New Zealand); vol. 1, p. 15.

p. 44 Lyell to Mantell, November 3, 1821. Turnbull Library, Wellington.

The full citation of Mantell's book is:

Mantell, G. A. *The fossils of the South Downs; or illustrations of the geology of Sussex.* London: Relfe, 1822.

p. 46 Dean's opinion about the first tooth is from:

Dean, D. R. "Gideon Mantell and the discovery of *Iguanodon.*" *Modern Geology* 18 (1993): 209–219.

p. 47 Mantell's comments on the tooth is from Mantell, 1851, p. 228.

p. 48 Mantell's quotation about Harper is from his transcribed journal, vol. 1, p. 88.

p. 49 Mantell's two quotations about the large ammonite are from his transcribed journal, vol. 1, pp.14–15.

Chapter 5

p. 54 Buckland's quotation about absence of evidence for water action is from:

Buckland, W. D. *Reliquiae Diluvianae.* London: Murray, 1824, p. 3.

p. 56 Buckland's comments about gelatine is from Buckland, 1824, p. 13.

The 121,000-year radiometric date is from:

McFarlane, D. A., and D. C. Ford. 1998. "The age of the Kirkdale Cave palaeofauna." *Cave and Karst Science* 25; *Transactions of the British Cave Research Association,* pp. 3–6.

p. 57 Buckland's quote about mud is from Buckland, 1824, pp. 10–12.

pp. 57–58 Buckland's quote about the Copley Medal is from a letter from Buckland to Cole, December 24, 1822. Archives, National Museum of Wales, Cardiff (84.20G.D165).

p. 59 Buckland's quote about the human female skeleton is from Buckland, 1824, p. 90.

p. 60 Buckland's quote about mixing of fossils is from Buckland, 1824, p. 85. His reference to ivory is from the same source, pp. 88–89.

Buckland's quote about the antiquity of the mammoth is from Buckland, 1824, p. 90. The radiometric date of 18,000 years is from:

Oakley, K. P. "The date of the 'Red Lady' of Paviland." *Antiquity* 42 (1968): 306–307.

p. 61 Buckland's quote about the exciseman is from a letter from Buckland to Cole, February 15, 1823. Archives, National Museum of Cardiff, Wales (84.20G.D167).

The student's poem is quoted in Gordon, 1894, p. 69.

Lyell's recollections about Buckland's exploits is contained in a letter from Lyell to Mantell, February 8, 1822, quoted in:

Lyell, K. *Life letters and journals of Sir Charles Lyell, Bart.* London: Murray, 1881.

MacEnery's quote is from:

p. 62 MacEnery, J., ed. *Cavern researches, discoveries of organic remains, and of British and Roman reliques, in the caves of Kent's Hole, Anstis Cove, Chudleigh and Berry Head.* Edited from the original manuscript notes by E. Vivian, London:, 1859. Quoted in:

Boylan, P. J. "Dean William Buckland, 1784–1856, a pioneer in cave science." *Studies in Speleology* 1 (1967): 237–253.

Chapter 6

pp. 67–69 Conybeare's quotations are from:

De la Beche, H. T., and W. D. Conybeare. "Notice of the discovery of a new fossil animal, forming a link between the Ichthyosaurus and Crocodile, together with general remarks on the osteology of the Ichthyosaurus." *Transactions of the Geological Society of London* 5 (1821): 559–594.

Although De la Beche appears as the senior author, the paper was written entirely by Conybeare.

pp. 71, 72 Conybeare's quotes about Buckland's surprise visit, the meeting, and the annual dinner are from a letter from Conybeare to De la Beche, March 4, 1824. Archives, National Museum of Wales, Cardiff (84.20G.D302).

pp. 71–72 Conybeare's quotes about plesiosaurs are from:

Conybeare, W. D. "On the discovery of an almost perfect skeleton of Plesiosaurus." *Transactions of the Geological Society* 1 (1824): 381–389.

His quote about when the plesiosaur skeleton was found by the "proprietor," and his reference to the necks of plesiosaurs are from the same source.

p. 77 Buckland's quote about the naming of *Megalosaurus* is from:

p. 77 (continued) Buckland, W. "Notice on the Megalosaurus or great fossil lizard of Stonesfield." *Transactions of the Geological Society of London* 1 (1824): 390–396.

The quotation about giants on Earth is from Genesis 6, verse 4. Buckland's reference to the Stonesfield lizard is from Buckland, 1824.

p. 79 The citation for Buckland's Bridgewater Treatises is:

Buckland, W. *Geology and mineralogy considered with reference to natural theology.* 2 vols. London: William Pickering, 1836.

Buckland's quote about marsupials is from Buckland, 1836, p. 74; his quote about the continuous chain is from the same source, p. 88.

Chapter 7

p. 81 Buckland to Mantell, March 3, 1824. Turnbull Library, Wellington.

p. 83 Buckland's comments about Mantell's fossils is from a letter from Buckland to Cuvier, June 2, 1824, quoted in Taquet 1982 (translated by S. Andrews).

Buckland's comment about Cuvier being puzzled about the large teeth is from a letter from Buckland to Mantell, June 2, 1824. Turnbull Library, Wellington.

Buckland's conversation with Miss Morland was recorded by Davies Gilbert and quoted in Gordon 1894, p. 91.

p. 84 Cuvier's comments about the teeth are in a letter from Cuvier to Mantell, June 20, 1824, quoted in Taquet 1982 (translated by S. Andrews).

Mantell's apology to Cuvier is contained in a letter from Mantell to Cuvier, July 9, 1824, quoted in Taquet 1982.

p. 85 Two entries in Mantell's journal, vol. 2, p. 143, for the fall of 1841 make reference to the perceived source of his back problems:

October 11. "Narrowly escaped severe injury from a fall occasioned by jumping out of my carriage to extricate my horse, the coachman having allowed the reins to become entangled. I fell with great violence, and the wheels just grazed my head but did not pass over me."

October 27. "Ill with symptoms of paralysis, arising from spinal disease brought on by over-exertion in stooping over my poor girl and Mr. Herbert in 1839 and 1840 . . ."

The reference to Mantell's posture is cited in:

Spokes, S. *Gideon Algernon Mantell, LL.D., F.R.C.S., F.R.S., surgeon and geologist.* London: John Bale, Sons and Danielsson, 1927, p. 7.

The quote about Clift is cited in:

Desmond, A. *The politics of evolution.* Chicago: University of Chicago Press, 1989, p. 419.

p. 86 Mantell's quote about the name *Iguanosaurus* is from a letter from Mantell to Cuvier, November 13, 1824, quoted in Taquet 1982.

p. 86 (continued) Buckland's comments about the name *Iguanosaurus* is from a letter from Buckland to Mantell, November 23, 1824. Turnbull Library, Wellington.

p. 87 Conybeare's comments are from a letter from Conybeare to Mantell, August 6 or 7, 1825. Turnbull Library, Wellington.

Mantell's quote about his immortality is from a journal entry, vol. 1, p. 144, for September 17, 1829.

Mantell's comment about the Royal Society is from a journal entry, vol. 1, p. 113, for December 22, 1825.

p. 88 Mantell's summary for the year is from the previous source (p. 114) for December 31, 1825.

His quote about the large fossil is from a journal entry, vol. 1, p. 133, for March 31, 1828.

pp. 89–91 Mantell's quote about the quarrymen is from a journal entry, vol. 2, p. 21, for December 12, 1830.

Mantell's quote, "ought to have been mine," is a journal entry, vol. 2, p. 36, for July 1, 1831. His quote about visiting Cuckfield is a journal entry for the following month, September 13, 1831; p. 39. His quote about visiting Mr. Trotter is from a journal entry, vol. 2, p. 40, for October 2, 1831.

Mantell's quote about royalty is a journal entry, vol. 2, p. 15, for October 22, 1830.

Mantell's quote about visitors is a journal entry, vol. 2, p. 2, for May 30, 1830. His quote about "first families" is a journal entry, vol. 1, p. 141, for December 31, 1828. His quote about moving to Brighton is a journal entry, vol. 2, pp. 26–27, for January 29, 1831.

Mantell's quote about the sermon is a journal entry, vol. 2, p. 10, for August 5, 1830. His quote about the assizes is a journal, vol. 2, pp. 21–23, for the end of the same year, December 31, 1830.

Mantell's comments about the fires were contained in journal entries, vol. 2, pp. 19–20, for November 19 and 25, 1830.

Chapter 8

pp. 94–95 Lyell's quote on Conybeare's Thames Valley paper is in a letter from Lyell to Mantell, April 1829, quoted in:

Lyell, K., ed. *Life letters and journals of Sir Charles Lyell, Bart.* London: Murray, 1881, pp. 252–253.

Lyell's quote about the "last discharge" is in a letter from Lyell to Mantell, June 7, 1829, source as previously cited, p. 253.

Mantell's quote about the saints in uproar is a journal entry, vol. 1, p. 151, for March 12, 1830.

p. 97 Darwin's quote on Lyell's views is from:

Darwin, F. 1887, ed., p. 73.

p. 98 Darwin's quote on following Lyell's example is from Darwin, F. 1887, ed., p. 83.

For a good recent discussion on the "Temple" of Serapis see:

Gould, S. J. "Pozzuoli's Pillars revisited." *Natural History* 5 (May 1999): 28–91 (the second installment of a two-part article; the first part appears in the April volume).

p. 100 Lyell's quote on organic life is from:

Lyell, C. *Principles of geology, being an attempt to explain the former changes of the Earth's surface, by reference to causes now in operation.* Vol. 1. London: Murray, 1830; facsimile of the first edition, Chicago: University of Chicago Press, 1990; p. 145.

p. 101 Lyell's quote on the reappearance of fossils is from Lyell, 1830, p. 123.

p. 102 Lyell's quote on centers of creation is from Lyell, 1830, p. 126.

Darwin's views on the same subject are from:

Darwin, C. *Journal of researches into the natural history and geology of the countries visited during the voyage of H.M.S."Beagle"round the world.* Minerva Library of Famous Books. London: Ward, Lock, 1845, p. 290. This, Darwin's first book, was first published in 1839.

p. 103 Darwin's quote on the great mystery is from Darwin, 1845, p. 275.

Chapter 9

p. 105 Data for dinosaur genera from:

Dodson, P. "What the fossil record of dinosaurs tells us." *Paleontological Society Special Publication* 7 (1994): 21–37.

p. 108 Mantell's quote is from a journal entry, vol. 2, p. 91, for August 14, 1834.

p. 111 Lecture advertised and reported in the *Brighton Gazette,* February 1834. Turnbull Library, Wellington.

p. 113 The *Lancet* was, and still is, an influential London medical journal. Report of the lecture quoted in:

Spokes 1927, pp. 83–84.

p. 114 The journal entry is from vol. 1, pp. 110–111. (Poem translated by Sarah Stewart.)

Chapter 10

p. 117 Owen's description of Hawkins is given in:

Owen, R. (Reverend). *The life of Richard Owen by his grandson.* Vol 1. London: Murray, 1894, p. 165. (Owen met Hawkins in 1839.)

Mantell's views on Hawkins, and his belief that he had a "few thousands" are contained in a letter from Mantell to Silliman, June 18, 1834. Silliman Family Papers, Manuscripts and Archives, Yale University Library.

p. 117 (continued) The reference to Hawkins's more modest inheritance was generously provided by Jehane Melluish, who has been conducting exhaustive researches on her distant relative. Hawkins's reference to himself are from page v of the preface to:

Hawkins, T. *Memoirs of Ichthyosauri and Plesiosauri, extinct monsters of the ancient Earth.* London: Relfe and Fletcher, 1834.

p. 118 The quote about giants is given in Hawkins,1834, p. 3.

p. 119 Extracts of Hawkins's conversations are given in Hawkins, 1834, pp. 33–34.

The quote about splitting the finder's fee between master and man is from:

Hawkins, T. *The book of the great sea-dragons, Ichthyosauri and Plesiosauri, Gedolim Taninim of Moses. Extinct monsters of the ancient Earth.* London: W. Pickering, 1840, p. 16.

Hawkins's views on the prehistoric habitat of the reptiles is given in Hawkins, 1834, p. 5.

p. 120 Hawkins's acceptance of the extinction of the "sea dragons" is given in the same source as previously.

Hawkins's views on the "unphilosophical" doctrine, and upon the "Creator's purpose" is given on page 12 of that source, and his views on Lamarck's heresy are from page 36 of the same volume.

pp. 122–123 Hawkins's recollection of the wet summer of 1833 and his adventures in Lyme are given in Hawkins, 1834, pp. 25–27.

His quote about his magic touch is given in the same source 1834, (p. 27.)

p. 124 Mantell's quote about Hawkins is from a journal entry, vol. 2, p. 63, for December 7, 1832.

pp. 125–126 Conybeare's remarks about Hawkins's book is from a letter from Conybeare to Buckland, August 4, 1834. National Museum of Wales, Cardiff (84.20G.D309).

Hawkins's quote about "Gedolim Taninim" is from Hawkins, 1840, p. 10, and his "cruel snake" quote is from p. 12 of the same source.

Hawkins's quote about mammary glands and the "horrid Brood" is from Hawkins, 1840, p. 18.

Hawkins's quote about the tail bend is from Hawkins, 1840, pp. 12–13.

The reference to the Earl of Kent is given in an unpublished autobiography of Joseph Clark III, in C. J. Clark Museum and Archives.

pp. 127–128 Hawkins's quote about the cholera is given in Hawkins, 1834, p. 41.

p. 130 The quote about the plaster man is contained in a letter from Hawkins to Mantell, January 29, 1833. Turnbull Library, Wellington.

The quote about the Italian artist is given in a letter from Mantell to Silliman, January 18, 1834. Manuscripts and archives, Yale University Library.

p. 131 The quotes by Hawkins are from Hawkins, 1834, p. 13.

p. 131 The Anning quote is from a letter from Anning to Murchison, October 11, 1833, quoted in:

Lang, W. D. "Three letters by Mary Anning, 'fossilist' of Lyme." *Proceedings of the Dorset Natural History and Archaeological Society* 66 (1945): 169–173.

Chapter 11

p. 134 Buckland to the trustees of the British Museum, July 7, 1834, quoted in:

Report from the Select Committee (of the House of Commons) on the condition, management, and affairs of the British Museum; together with the minutes of evidence, appendix, and index. London: House of Commons (located in the House of Lords Library, available in the House of Commons Library), 1836, p. 440.

Mantell's quote about the ichthyosaur is from a journal entry, vol. 2, p. 63, for December 7, 1832. His quotation about the evaluation is from the same source, p. 89, for July 12, 1834.

p. 135 Buckland's first letter, to the trustees of the British Museum on July 12, 1834, quoted in the Select Committee Report, p. 441.

Buckland's second letter, to Hawkins on July 9, 1834, is quoted in:

Charlesworth, E. "Appendix." *Magazine of Natural History, New Series* 4 (1840): 29. (Also known as the *Annals and Magazine of Natural History*.)

Hawkins to Buckland, July 9, 1834, quoted in Charlesworth 1840, pp. 35–36.

p. 136 Hawkins to Mantell, July 2, 1834. Turnbull Library, Wellington.

pp. 138–139 The quote about the size of the large ichthyosaur is given in Hawkins, 1834, plate 3.

König to Buckland, January 20, 1835. London: Archives, Natural History Museum. By permission of the trustees of the National History Museum.

König's formal report to the trustees, February 12, 1835, quoted the Select Committee Report, 1836, p. 441.

pp. 140–141 Buckland to Forshall, February 12, 1835, quoted in the Select Committee Report, 1836, pp. 443–444.

pp. 142–143 Grant's testimony, March 24, 1836, quoted in the Select Committee Report, 1836, p. 133.

Hawkins to the trustees of the British Museum, 15 February, 1841, published in:

Hawkins, T. "Statement relative to the British Museum." London: published by the author, 1848.

p. 144 Mantell to Silliman, September 25, 1835. Silliman Family Papers, Manuscript and Archives, Yale University Library.

pp. 145–146 The strawberry incident is recounted in the unpublished Clark manuscript.

pp. 145–146 (continued) The reference to Charlesworth's brief curatorship at the British Museum is given in:

Cleevely, R. J. *World Palaeontological Collections.* London: British Museum (Natural History) and Mansell Publishing, 1983.

Hawkins's first letter to Young, dated December 29, 1838, and the second one, dated January 4, 1839, are both quoted in Charlesworth, 1840, p. 15.

Hawkins's letter to Charlesworth's solicitor is quoted in Charlesworth, 1840, pp. 17–18.

Lyell's quote about Charlesworth's pending ruin is quoted in Charlesworth, 1840, p. 19.

The quotations about Buckland's willingness to be a mediator and the terms of settlement are contained in a letter from Lyell to Charlesworth. The letter is undated, but was received on February 17, 1839. Quoted in Charlesworth, 1840, pp. 19–20.

pp. 147–148 Forshall to Charlesworth, February 18, 1839. Quoted in Charlesworth, 1840, p. 20.

The document that fell into Charlesworth's hands is quoted in Charlesworth, 1840, p. 21.

Charlesworth's withering conclusion is quoted in Charlesworth, 1840, pp. 43–44.

Chapter 12

pp. 152, 154 The extract from Caroline's diary is quoted in Owen, R. (Reverend), 1894, vol. 1, p. 109.

pp. 154–155 The quotes from Owen's lecture are from:

Sloan, P. R., ed. *The Hunterian lectures in comparative anatomy May-June, 1837 Richard Owen.* London: Natural History Museum Publications, 1992, p. 79.

Owen was probably alluding to the "transcendental anatomy" of the German anatomist Johann Spix, and of the French embryologist Etienne Serres. See the introductory chapter of this reference.

Owen's quote about the abuse of homology is from Sloan, 1992, p. 112.

p. 156 The quote from the *London Medical and Surgical Journal* is given in Sloan, 1992, p. 67.

The two quotes from Caroline's diary are from Owen, R. (Reverend), 1894, vol. 1, pp. 109, 112, respectively.

p. 157 Owen's quote about transmutation is given in:

Sloan, 1992, p. 192.

The citation for Owen's book is:

Owen, R. *Instances of the power of God as manifested in his animal creation.* London: Longman, Green, 1864.

p. 157 (continued) Owen's quote on progressive development and his quote about extinct species are given in Sloan, 1992, pp. 223, 222, respectively.

p. 158 Caroline's three diary quotes are given in Owen, R. (Reverend), 1894, vol. 1, pp. 94, 97, and 98, respectively.

Caroline's quote about the rhinoceros are contained in the previous reference, p. 121.

p. 159 Lyell to Darwin, July 26, 1836, quoted in previous reference, p. 102.

Darwin's comment about smoky London is contained in a letter he wrote to Jenyns, April 10, 1837, quoted in:Darwin, F. 1887, ed., vol. 1, p. 282.

p. 160 Darwin's comment about Owen's willingness to dissect pickled specimens is contained in a letter he wrote to Henslow, October 1836, quoted in:

Burkhardt, F., and S. Smith, eds. *The correspondence of Charles Darwin*. Vol. 1. Cambridge: Cambridge University Press, 1985, p. 512.

Darwin's comment on Lyell's kindness is contained in a letter he wrote to his cousin, Fox, November 6, 1836. Quoted in Darwin, F. 1887, ed., vol. 1, p. 277.

Darwin's comment about reading papers to the Geological Society is contained in a letter he wrote to Fox, July 1837, quoted in the previous reference, p. 280.

Lyell's comment on Darwin's coral island theory is contained in a letter he wrote to Herschel, May 24, 1837, quoted in:

Lyell, K., ed. *Life Letters and Journals of Sir Charles Lyell, Bart*. Vol. 2. London: Murray, 1881, p. 12.

Darwin's comment of having a capital friend in Lyell is contained in a letter he wrote to Jenyns, April 10, 1837, quoted in Darwin, F., ed., 1887, vol. 1, p. 282.

p. 161 Caroline's comment on the Darwin visit is quoted in Owen, R. (Reverend), 1894, vol. 1, p. 108.

Darwin's comment on making out the remains of animals is quoted in a letter he wrote to Lyell, July 30, 1837, quoted in:

Burkhardt, F., ed. *Charles Darwin's letters. A selection 1825–1859*. Cambridge: Cambridge University Press, 1996, p. 59.

p. 162 Darwin's first comment on the cause of death is quoted in the previous letter he wrote to Lyell, July 30, 1837. His second comment, asking what has exterminated so many animals, is quoted from:

Darwin, C. *Journal of researches into the geology and natural history of the various countries visited by H.M.S. Beagle*. Facsimile reprinting of the first edition, published in 1952 by Hafner Publishing Company, New York.

p. 164 Owen's concern for Lord Eniskillen's safety is quoted from Owen, R. (Reverend), 1894, vol. 1, p. 171.

Owen's account of his trip to the West Country is quoted in the previous reference, pp. 165–166.

Chapter 13

p. 166 The quotes about the landslide are from Gordon, 1894, pp. 175–176.

The glacial hypothesis, which predates De Charpentier, was discussed by Playfair as early as 1802:

Playfair, J. *Illustrations of the Huttonian theory of the earth.* Edinburgh: William Creech; London: Cadell and Davies, 1802.

Mrs. Buckland's comments are quoted in Gordon, 1894, p. 141.

Murchison's comment is quoted in:

Marcou, J. *Life, letters and works of Louis Agassiz.* New York: Macmillan, 1896, p. 167.

Buckland's letter to De la Beche about glaciers was written November 19, 1840. Archives, National Museum of Wales, Cardiff (84.20G.D189).

Conybeare's letter to Buckland about being frostbitten was written on December 15, 1840. Archives, National Museum of Wales, Cardiff (84.20G.D311).

p. 172 Conybeare's second letter to Buckland about glaciation was dated November 19, but without the year [1840 or 1841]. Archives, National Museum of Wales, Cardiff.

Conybeare raised the question about the flood in:

Conybeare, W. D. *The Philosophical Magazine* also known as *Annals of Chemistry, Astronomy, Natural History and General Science*, 9 (1831), p. 190.

Agassiz's reference to God's plough was quoted in:

Lurie, E. *Louis Agassiz: A life in science.* Chicago: University of Chicago Press, 1960, p. 98.

Chapter 14

p. 175 Torrens, H. "When did the dinosaur get its name?" *New Scientist* 134 (1992): 40–44.

Owen, R. "Report on British fossil reptiles." Part 2. *Report of the British Association for the Advancement of Science, Plymouth* 11 (1842): 60–204.

This report, specifically Mantell's personal annotated copy, has recently been republished:

Dean, D. R., ed. *The first "Dinosaur" Book. Richard Owen on British fossil reptiles (1842).* Scholars' Facsimiles and Reprints. New York: Delmar, 1999. Two of Mantell's annotations bear the expletive "fudge," where he disagrees with Owen's identifications of specific bones of *Iguanodon.*

pp. 176–181 All of Owen's comments are quoted from Owen, R., 1842.

Torrens, 1992, has shown that Owen revised his own length estimates downward during the interval between giving his presentation at Plymouth and writing the published report.

p. 184 Owen's anonymous review of the *Origin* was published in:

Edinburgh Review 3 (1860): 487–532.

p. 185 Darwin's comment on being quiet in London is contained in a letter he wrote to Fox, October 1839, quoted in Darwin, F., ed., 1887, vol. 1, p. 299.

Darwin's comment on the vile smoky place is contained in a letter he wrote to Jenyns, April 10, 1837, quoted in Darwin, F., ed., 1887, vol. 1, p. 282.

Darwin's longing for pure air is contained in a letter he wrote to Fox, September 28, 1841, quoted in Burkhardt, F., and S. Smith, eds. *The correspondence of Charles Darwin.* Cambridge: Cambridge University Press, 1986, vol. 2, p. 305.

The reference to Darwin's height is given in Darwin, F., ed., 1887, vol. 1, p. 109.

The reference to Darwin's weight is given in footnote 5 to a letter Darwin wrote to Emma Darwin, April 5, 1840, given in: Burkhardt and Smith, eds., 1986, p. 263.

Darwin's comment of the country air working magic is contained in a letter he wrote to Jenyns, June 24, 1841, quoted in previous reference, p. 292.

Darwin's comment about seeing scarcely anybody is contained in a letter he wrote to Fox, January 25, 1841, quoted in previous reference, p. 279.

Darwin's comments to Fox on his work on varieties and obtaining carcasses is contained in the previously cited letter.

p. 186 The prime minister's quote is taken from a letter Peel wrote to Owen, November 1, 1842, cited in Owen, R. (Reverend), 1894, vol. 1, pp. 203–204.

Whewell's comment is taken from a letter he wrote to Owen, quoted in the same reference as previously, pp. 204–205.

p. 187 The quote about Owen's distress at the misery is given in the same reference as previously, p. 217.

The quote about Owen finding his way to Whitechapel is given on the same page as before.

Caroline's quote about declining the knighthood is given in the same reference as before, p. 262.

p. 188 Sedgwick's comment is given in the introduction (p. 31) to the 1969 reprinting of *Vestiges:*

Chambers, R. *Vestiges of the Natural History of Creation, with an introduction by Gavin de Beer.* New York: Humanities Press, 1969.

Darwin's comments on *Vestiges* is contained in a letter he wrote to Hooker, 1844 [?] quoted in Darwin, F., ed., 1887, vol. 1, p. 333.

Other quotes on p. 188 and all on p. 189 from *Vestiges.*

p. 190 Sedgwick's concerns are quoted from:

Desmond, A., and J. Moore. *Darwin.* London: Michael Joseph, 1991, p. 321.

Murchison's quote is included in a letter he wrote to Owen, April 2, 1845, cited in Owen, R. (Reverend), 1894, vol. 1, p. 254.

p. 190 (continued) Owen's response to the author of *Vestiges* is quoted in the previous reference, p. 249.

Chapter 15

p. 194 Buckland to Faraday, 13 June, 1849, quoted in Gordon 1894, pp. 223–224.

p. 195 Lyell to his sister, February 27, 1848, quoted in Lyell, K., ed., 1881, vol. 2, p. 139.

p. 196 The reference to Queen Victoria is from an entry for April 8, 1848, in Mantell's journal, vol. 3, p. 157.

pp. 197–198 Mrs. Buckland's comment about her husband's health is contained in a letter she wrote to De la Beche, March 15, 1850. Archives, National Museum of Wales, Cardiff (84.20G.D142).

p. 199 The discovery that Hawkins married and adopted a son is owing to the untiring researches of his distant relative, Jehane Melluish, and is reported in:

Taylor, M. A. "Hawkins." In *New Dictionary of National Biography.* Oxford: Oxford University Press, 2000.

p. 201 De la Beche, H. T. "Anniversary address of the President." *Quarterly Journal of the Geological Society of London* 4 (1848): xxiv–xxv.

Chapter 16

p. 204 The Pinney comment is quoted in:

Lang, W. D. "Mary Anning and Anna Maria Pinney." *Proceedings of the Dorset Natural History and Archaeological Society* 76 (1956): 146–152.

p. 206 McGowan, C. "A putative ancestor for the swordfish-like ichthyosaur *Eurhinosaurus*." *Nature* 322 (1986): 454–456.

pp. 208–209 The new species *L. solei* was originally referred to the genus *Leptopterygius,* but this is no longer a valid ichthyosaur name and the appropriate genus is *Leptonectes.* See:

McGowan, C. "A new species of large, long-snouted ichthyosaur from the English lower Lias." *Canadian Journal of Earth Sciences* 30 (1993): 1197–1204.

McGowan, C. "The taxonomic status of *Leptopterygius* Huene, 1922 (Reptilia: Ichthyosauria)." *Canadian Journal of Earth Sciences* 33 (1996): 439–443.

p. 210 The sinking of the name *T. risor* is discussed in:

McGowan, C. "*Temnodontosaurus risor* is a juvenile of *T. platyodon* (Reptilia: Ichthyosauria)." *Journal of Vertebrate Paleontology* 14 (1994a): 472–479.

p. 212 As the original type specimen of the species *Leptonectes tenuirostris* had been lost, I designated another specimen as the neotype (the replacement type), back in 1974. I chose the most perfect specimen—the one with the artificial matrix—before I recognized the problem regarding its authenticity. See:

p. 212 (continued) McGowan, C. "A revision of the latipinnate ichthyosaurs of the Lower Jurassic of England (Reptilia: Ichthyosauria)." *Life Sciences Contributions Royal Ontario Museum* 100 (1974): 1–30.

McGowan, C. "*Leptopterygius tenuirostris* and other long-snouted ichthyosaurs from the English Lower Lias." *Palaeontology* 32 (1989): 409–427.

McGowan, C. "Problematic ichthyosaurs from southwest England: A question of authenticity." *Journal of Vertebrate Paleontology* 10 (1990): 72–79.

pp. 215–217 For further information about the unknown giant ichthyosaur from England, see:

McGowan, C. "Giant ichthyosaurs of the Early Jurassic." *Canadian Journal of Earth Sciences* 33 (1996): 1011–1021.

p. 217 For more information on the preservation of microscopic structures see:

Martill, D. M., and D. M. Unwin. "Exceptionally well-preserved pterosaur wing membrane from the Cretaceous of Brazil." *Nature* 340 (1989): 138–140.

Wyckoff, R. W. G. "Trace elements and organic constituents in fossil bones and teeth." *Proceedings of the North American Paleontological Convention for 1969*. Lawrence, Kans.: Allen Press, 1971.

Further Reading

Chapter 2

Tickell, C. *Mary Anning of Lyme Regis.* Lyme Regis, England: Philpot Museum, 1996.

Torrens, H. S. "Collections and Collectors of Note: 28 Colonel Birch (c. 1768–1829)." *Geological Curator* 2:7 (1979): 405–412.

______. "Collections and Collectors of Note: 28 Colonel Birch (c. 1768–1829)." *Geological Curator* 2:9, 10 (1980): 561–562.

______. "Mary Anning (1799–1847) of Lyme; 'The Greatest Fossilist the World Ever Knew.'" *British Journal of the History of Science* 28 (1995): 257–284.

Chapter 3

Buckland, W. *Vindiciae Geologicae; or the connexion of geology with religion explained.* Oxford: Oxford University Press, 1820.

Edmonds, J. M. "*Vindiciae Geologicae*, published in 1820; the inaugural lecture of William Buckland." *Archives of Natural History* 18 (1991): 255–268.

Edmonds, J. M., and J. A. Douglas. "William Buckland, F.R.S. (1784–1856) and an Oxford geological lecture, 1823." *Notes and Records, Royal Society* 30 (1976): 141–167.

Gordon, E. O. *The life and correspondence of William Buckland, D.D., F.R.S.* London: Murray, 1894.

Rudwick, M. J. S. *The meaning of fossils. Episodes in the history of palaeontology.* London: MacDonald, 1972.

______. *Georges Cuvier, Fossil Bones and Geological Catastrophes. New Translations and Interpretations of the Primary Texts.* Chicago: University of Chicago Press, 1997.

Rupke, N. A. *The great chain of history.* Oxford: Clarendon Press, 1983.

Taquet, P. "Cuvier-Buckland-Mantell et les Dinosaures." In *Actes du Symposium Paléontologique G. Cuvier, Montbeliard,* ed. E. Buffetaut, J. M. Mazin, and E. Salmon, 1983, pp. 475–491.

Chapter 4

Bettmann, O. L., and P. S. Hench. *A pictorial history of medicine.* Springfield, State: Charles C. Thomas, 1956.

Briggs, A. *The age of improvement 1783–1867.* London: Longman, 1959.

Cleeveley, R. J., and S. D. Chapman. "The accumulation and disposal of Gideon Mantell's Fossil Collections and their role in the history of British palaeontology." *Archives of Natural History* 19 (1992): 307–364.

Curwen, E. C., ed. *The journal of Gideon Mantell, surgeon and geologist.* London: Oxford University Press, 1940.

Dean, D. R. "Gideon Mantell and the discovery of *Iguanodon*." *Modern Geology* 18 (1993): 209–219.

———. *Gideon Mantell and the discovery of dinosaurs.* Cambridge: Cambridge University Press, 1999.

Lawrence, C., ed. *Medical theory, surgical practice.* London: Routledge, 1992.

Magner, L. N. *A history of medicine.* New York: Marcel Dekker, 1992.

Smith, F. B. *The people's health 1830–1910.* London: Croom Helm, [DATE missing].

Spokes, S. *Gideon Algernon Mantell.* London: John Bale Sons and Danielsson, 1927.

Trevelyan, G. M. *British history in the nineteenth century (1782–1901).* London: Longmans, Green, 1927.

Chapter 5

Kurtén, B. *Pleistocene Mammals of Europe.* London: Weidenfeld and Nicolson, 1968.

Sutcliffe, A. J. *On the Track of Ice Age Mammals.* Cambridge: Harvard University Press, 1985.

Chapter 7

Rudwick, M. J. S. *The meaning of fossils.* London: MacDonald, 1972.

Spokes, S. "A case of circumstantial evidence." *Sussex County Magazine* 11 (1937): 118–122.

MS-papers 0083–123A in the Turnbull Library, being a collection of newspaper articles and letters of Mantell relevant to the trial of Hannah Russell.

Chapter 12

Appel, T. A. *The Cuvier-Geoffroy debate.* New York: Oxford University Press, 1987.

Desmond, A. *The politics of evolution.* Chicago: University of Chicago Press, 1989.

Desmond, A., and J. Moore. *Darwin.* London: Michael Joseph, 1991.

Herbert, S., ed. *The red notebook of Charles Darwin.* London: British Museum of Natural History, 1980.

Gruber, J. W., and J. C. Thackray. *Richard Owen Commemoration.* London: Natural History Museum Publications, 1992.

Rupke, N. A. *Richard Owen Victorian naturalist.* New Haven: Yale University Press, 1994.

Sulloway, F. J. "Darwin's Conversion: The Beagle voyage and its aftermath." *Journal of the History of Biology* 15 (1982): 325–396.

Chapter 14

Owen, R. "On the *Archaeopteryx* of Von Meyer with a description of the fossil remains of a long-tailed species, from the lithographic stone of solenhofen." *Transactions of the Royal Society* 153 (1864): 33–47.

Padian, K. "The rehabilitation of Sir Richard Owen." *Bioscience* 47 (1997): 446–453.

Sources of Illustrations and Credits

p. 4 Skeleton: Kingsley, J. S., ed. The standard natural history. Volume 5, Mammals. Boston: S.E. Cassino, 1884.

Teeth: Owen, R. Palaeontology or a systematic summary of extinct animals and their geological relations. Edinburgh: Adam and Charles Black, 1861.

p. 13 By kind permission of John Fowles

p. 14 © Natural History Museum

p. 15 By kind permission of Mr. Roderick Gordon and Mrs. Diana Harman.

p. 17 By kind permission of John Fowles.

p. 21 Hawkins, T. Memoirs of Ichthyosauri and Plesiosauri, extinct monsters of the ancient Earth. London: Relfe and Fletcher, 1834.

p. 25 Courtesy Oxford University Museum of Natural History.

p. 32 Courtesy Oxford University Museum of Natural History.

p. 35 Woodward, H. B. The history of the Geological Society of London. London: Geological Society, 1907. Reproduced by permission of the Geological Society of London, owners of the publication right therein.

p. 39 Negative no. 2800. Courtesy Department of Library Services, American Museum of Natural History.

p. 45 © Natural History Museum.

p. 46 Nicholson, H. N. A manual of palaeontology. Volume 2. Edinburgh: William Blackwood and sons, 1879.

p. 53 Buckland, W. D. Reliquiae Diluvianae. London: Murray, 1824.

p. 55 Buckland, 1824.

p. 59 Buckland, 1824.

p. 68 Conybeare, W. D. "Additional notices on the fossil genera Ichthyosaurus and Plesiosaurus." Transactions of the Geological Society of London 1 (1822).

p. 69 Conybeare, W. D. "On the discovery of an almost perfect skeleton of Plesiosaurus." Transactions of the Geological Society of London 1 (1824).

p. 70 © National Museum of Wales.

p. 73 Richardson, G. F. Geology for beginners. London: Baillière, 1842.

p. 76 Buckland, W. "Notice on the Megalosaurus or great fossil lizard of Stonesfield." Transactions of the Geological Society of London 1 (1824).

p. 79 Mantell, G. A. Petrifactions and their teachings. London: H. G. Bohn, 1851.

p. 82 Mantell, G. A. The wonders of geology. London: Relfe and Fletcher, 1839.

p. 86 Mantell, 1851.

p. 94 Woodward, H. B. History of Geology. New York: G. P. Putnam's Sons, 1911.

p. 96 Science & Society Picture Library, Science Museum, London (NRP/RLO/COOO189A)

p. 99 Lyell, C. Principles of geology, volume 1. London: Murray, 1830.

p. 106 Mantell, 1839.

p. 107 Drawing by Ian Morrison, in possession of author.

p. 109 Mantell, 1851.

p. 110 Mantell, 1839.

p. 112 Courtesy Alexander Turnbull Library, National Library of New Zealand, Te Puna Mātauranga o Aotearoa, Wellington, NZ. Reference number E–330-f–001, drawing by George Scharf (1788–1860).

p. 124 Hawkins, 1834

p. 127 Courtesy Somerset County Council Museums Service.

p. 129 Hawkins, 1834.

p. 138 Photograph taken by author.

p. 139 Hawkins, 1834.

p. 150 Owen, R (Reverend). The life of Richard Owen by his grandson. Volume 1. London: Murray, 1894.

p. 151 Illustrated London News, October 4, 1845.

p. 153 Courtesy Library, Royal College of Surgeons of England.

p. 162 Owen, 1861.

p. 163 Owen, 1861.

p. 167 Gordon, E. O. The life and Correspondence of William Buckland. London: Murray, 1894.

p. 169 Cockburn, J. Swiss scenery from drawings by Major Cockburn. London: Rodwell & Martin, 1820.

p. 176 © Natural History Museum.

p. 178 © Natural History Museum.

p. 181 Top: Hawkins, 1834. Bottom: drawing by Julian Mulock, in possession of author.

p. 183 Top: drawing by Julian Mulock, in possession of author. Bottom: Sedgwick, A. A students text-book of zoology. London: Swan Sonnenschein and Co., 1905.

p. 199 Owen, R (Reverend), volume 2, 1894.

p. 208 By kind permission of Roger Clark.

p. 209 Photographs by author.

p. 212 By kind permission of Chris Pamplin.

p. 213 Photograph by author.

Index